I0590164

Communications After AD2000

Technology in the Third Millennium

1. **The Treatment and Handling of Wastes**
   Edited by A.D. Bradshaw, Sir Richard Southwood and Sir Frederick Warner
2. **A Global Strategy for Housing in the Third Millennium**
   Edited by W.A. Allen, R.G. Courtney, E. Happold and Sir Alan Muir Wood
3. **Communications After AD2000**
   Edited by D.E.N. Davies, C. Hilsum and A.W. Rudge

Communications After AD2000

Edited by
D.E.N. Davies,
C. Hilsum and
A.W. Rudge

Published by Chapman & Hall for The Royal Society

Published by Chapman & Hall, 2–6 Boundary Row, London SE1 8HN

Chapman & Hall, 2–6 Boundary Row, London SE1 8HN, UK

Blackie Academic & Professional, Wester Cleddens Road, Bishopbriggs, Glasgow G64 2NZ, UK

Chapman & Hall Inc., 29 West 35th Street, New York NY10001, USA

Chapman & Hall Japan, Thomson Publishing Japan, Hirakawacho Nemoto Building, 6F, 1-7-11 Hirakawa-cho, Chiyoda-ku, Tokyo 102, Japan

Chapman & Hall Australia, Thomas Nelson Australia, 102 Dodds Street, South Melbourne, Victoria 3205, Australia

Chapman & Hall India, R. Seshadri, 32 Second Main Road, CIT East, Madras 600 035, India

First edition 1993

Typeset in 10/12 pt Times by Excel Typesetters Company
Printed in Great Britain at the University Press, Cambridge

ISBN 0 412 49550 3

A catalogue record for this book is available from the British Library

Library of Congress Cataloging-in-Publication data
Communications after AD2000 / edited by D.E.N. Davies, C. Hilsum, and A.W. Rudge. – 1st ed.
p. cm. – (Technology in the third millenium; 3)
Based on a meeting of the Royal Society.
Includes bibliographical references and index.
ISBN 0–412–49550–3 (alk. paper)
1. Telecommunication – Congresses. I. Davies, D.E.N. II. Hilsum, Cyril. III. Rudge, A.W. (Alan W.) IV. Royal Society (Great Britain) V. Series.
TK5101.C6576 1993
384 – dc20 92–46391
CIP

Contents

Contributors

Sir Bryan Carsberg
Director General
Oftel
Export House
50 Ludgate Hill
London
EC4M 7JJ

Joseph De Feo
Chief Executive, Information Technology and Service Businesses Division
Barclays Bank PLC
Johnson Smirke Building
4 Royal Mint Court
London EC3N 4HJ

Dr Gary J. Handler
Vice President
Information Networking Service
Bellcore
331 Newman Springs Road
Redbank
NJ 07701-7030
USA

Professor C.A.R. Hoare, FRS
Computing Laboratory
Oxford University
8-11 Keble Road
Oxford OX1 3QD

Dr Nigel W. Horne
KPMG Management Consulting
P.O. Box 695
8 Salisbury Square
London EC4Y 8BB

Dr. Andy Hopper
University of Cambridge
Computer Laboratory
New Museums Site
Pembroke St.
Cambridge CB2 3QG
and Olivetti Research Limited
Old Addenbrookes Site
24a Trumpington Street
Cambridge CB2 1QA

Dr. John S. Mayo
President
AT&T Bell Laboratories
101 John F. Kennedy Parkway
P.O. Box 1101
Short Hills
NJ 07078-0996
USA

John E. Midwinter FEng, FRS
Pender Professor and Head of Department
Department of electronic and
Electrical Engineering
University College London
Torrington Place
London WC1E 7JE

Professor Ian Miles
PREST
Programme of Policy Research in Engineering Science & Technology
The University
Manchester M13 9PL

Dr T.R. Rowbotham
Director, Network Technology
BT Development and Procurement
BT Laboratories
Martlesham Heath
Ipswich 1PS 7RE

Mr C.P. Sandbank
British Broadcasting Corporation
White City
201 Wood Lane
London W12 7TS

Professor Raymond Steele
Professor of Communication
Department of Electronics and Computer Science
University of Southampton
Southampton SO9 5NH

Mr G. Tenzer
Member of the Management Board
Deutsche Bundespost Telekom
Generaldirektion
Postfach 2000, S300 Bonn 1.
Germany

Dr Iwao Toda
Executive Vice President
Nippon Telegraph and Telephone Corporation
Yamato Life Insurance Building, 10th floor
1–7, Uchisaiwai-cho 1-chome Chiyoda-Ku
Tokyo 100
Japan

Mr Iain D.T. Vallance
Chairman
British Telecommunications plc
BT Centre
81 Newgate Street
London ECIA TAJ

Mr Martin Ward
Technical Director, GPT Limited
New Century Park
PO Box 53
Coventry CV3 IHJ

Preface

The Royal Society has initiated a series of meetings to discuss the effect advances in technology will have on our way of life in the next century. The two previous meetings have been concerned with housing and waste treatment. The subject of the third meeting, communications, is no less critical to life, but it offers particular problems and uncertainties, especially in the forecasting of future trends. Indeed, some have doubted if there can be profitable debate on long-term development in such a fast-moving field. The importance of the topic justifies an attempt, and the reader will judge whether the authors have met the challenge.

Communications today bears little resemblance to that of the 1970s. Then we knew about satellites and optical fibres, and we had seen lasers and silicon chips, but most of us could never imagine the potential of the new technologies within our grasp. We had also not assessed the thirst of the population for more and better ways of talking and writing to each other. It was the combination of market need and technical capability that created the communications revolution.

The very nature and magnitude of that revolution gives us confidence that the next wave of change will be evolutionary, and, to some degree, predictable. There is so much to do in exploiting and refining what we now have that the future will be set more by what we want and can afford than by any miracles of invention. Nevertheless, the impact on life will be tremendous.

Easy and cheap communications will certainly improve our methods of working and playing, but there is a deeper significance. Salvador de Madariaga wrote 'Liberty of thought means liberty to communicate one's thought'. The inverse, liberty to communicate leading to liberty of thought, may have been a significant factor in the collapse of totalitarian government in a number of countries. This theme was developed at the meeting, with a general assertion that national prosperity and the standard of living depended on communications, so that governments have a part to play in easing the way to progress and encouraging international cooperation.

This was a unique opportunity for the leaders of the networks to explain their forward strategies and for the technical experts to explain how science and engineering would help them achieve their goals. It was an important occasion, and this book encapsulates the proceedings. Here the scene is set, with the way forward marked out. In a field of critical importance to all of us, such a lasting record will be in great demand.

Cyril Hilsum CBE, FEng., FRS

1

Global strategies

I.D.T. Vallance, Chairman, BT

My topic, global strategies, sounds almost mundane by comparison to Communications after AD2000 in the context of Technology in the Third Millennium. Such ringing phrases are irresistible. You don't have to be an armchair clairvoyant or a Nostradamus *manqué* to feel the lure of the future. Indeed, picturing it is a key activity for all but the most parochial and complacent organizations. So, before I consider the global strategy, I'd like to have a shot at predicting the future of telecommunications.

In the last few years, the telecommunications industry has begun to merge with the communications services sector which, in turn, has converged with information services and entertainment services to produce the hydra-headed information technology or IT industry. An exciting combinaton of customer demand, increasing competition and advancing technology has created a fast-moving industry, which is constantly transforming itself in the development of new areas of business and which is constrained only by regulation.

Thus far, advances in technology have been the major force for change (although there is no reason why this should inevitably continue to be the case). The price performance of hardware has improved by about 30% per annum for the past ten years and is forecast to improve at the same rate over the next ten. We can now get more bangs per buck, more megabytes per dollar than we could ever have imagined a few years ago; it is equally difficult to imagine the implications of these continuing and dramatic changes ten years out.

A recent special issue of *Scientific American* had this to say about the dawning of the information age:

> The authors of this issue share a hopeful vision of a future built on an information infrastructure that will enrich our lives by relieving us

Communications After AD2000. Edited by D.E.N. Davies, C. Hilsum and A.W. Rudge. Published in 1993 by Chapman & Hall, London, for The Royal Society. ISBN 0 412 49550 3

> of mundane tasks, by improving the ways we live, learn and work and by unlocking new personal and social freedoms . . . A growing opportunity has attained critical mass as a result of a twofold serendipity: dramatic improvements in the cost performance ratios of computers and of communications technology . . . This relentless compounding of capabilities has transformed a faint promise of synergy into an immense and real potential.

By the end of the 1990s, advances in Information Technology will allow us to provide services to our customers anywhere, anyway, and anytime they want it. This is a very challenging vision, but one that is underpinned by two unstoppable forces that are with us now – digitalization and the increasing level of software control, or intelligence, in the network. Today, voice, video, image, character, and data are still treated more or less separately. But to a digital network these elements are all just 0s and 1s, on and off, which means they are capable of being re-formed in any way that the customer wants, whether it be speech into text or data into images. This malleability of information will allow the technology to shrink-wrap itself around the individual and be unique to a person's needs and desires. You can see the beginnings of this trend in the world of personal computers. Customers take standard hardware platforms, standard operating systems, and off-the-shelf application programs. In no time at all the machine is customized to the needs of the person who uses it. Telecommunications services are going the same way. Already, virtual private networks allow businesses to control their service levels and network configuration.

The information intensive individual will likewise demand access to the services that he or she wants, anywhere and any time – and that means being allowed to take an active role in specifying, developing and managing the system he or she requires.

Communications will be associated with people rather than physical locations. The science fiction dream of everyone being issued with a unique communications identity at birth and being contactable anywhere on earth, throughout their lives, is not that far away. The Personal Communications initiatives being developed by BT aim to do just that. Some of you might consider the idea to be more of a nightmare than a dream, but there will always be an 'off' button!

Businesses in the global networked society of the near future will be more complex, more interrelated and more dynamic. The multinationals of the future, as the multinationals of here and now have discovered, will require information technology supplied on a global basis. In their case, big will be beautiful. Those who wish to win their business will have to operate globally, and refuse to be hamstrung by national boundaries.

BT's vision of becoming the most successful worldwide telecommunications group is a clear and direct response to this challenge.

Going global is emphatically not the same thing as going multidomestic. It requires new ways of thinking about business in which geographical and organizational barriers will be increasingly an irrelevance. The scale of the exercise is so vast that old notions of competition are being abandoned. No one company will ever own enough of the global communications infrastructure to ensure single-handed delivery around the globe. Meeting customers' needs will require more partnerships and joint ventures. Companies will continue both to compete, where that is in the interests of customers, and to learn to co-operate if that is what is needed. Paradoxically, in the global networked society your best customers can also be your most active competitors.

In the 21st century, geographically dispersed organizations will be able to function as coherent wholes. Picture the scene: directors in boardrooms in several countries, one in a plane, another on a train, another working from home, all participating in a three-dimensional video conference, seeing and talking to one another as though they were in the same room. The basic technology is already here.

And there, for the moment, we should leave the future. To bring us down to earth a little, the point that I must stress is that although our projections are based on the best information available, there is always the nagging possibility that the future will be different in fundamental ways from our best guesses. Speculation is very necessary, but we should never lose sight of the fact that it is an inexact science. Indeed, if predicting the future was as easy as that, we probably wouldn't be sitting here today.

The only thing we can say with certainty is that the future is essentially unknowable and elusive.

It follows that you cannot base a successful business strategy on a detailed picture of the future, whether that picture is derived from economic forecasting or any other form of divination. The world in which we operate is not fixed and determined. Building a strategy on the conviction that you know exactly what the world will be like in five or ten years' time is no more and no less scientific than backing every favourite on the card at the races!

The only viable strategy is based on imaginative commitment to the customer and speed of reaction to changing circumstances and priorities. We cannot say exactly what the world will be like but we can put ourselves in a position to compete, whatever it is like. Successful companies will be those that listen to their customers, ones that are willing to change as the times change and are better at changing than their competitors. That is the only form of long-term sustainable strategic advantage. The key to strategic success is precisely not being burdened with a too detailed view of the future.

To ensure that we really do achieve the necessary degree of readiness, BT's strategy is based not on our powers of divination but on four first principles, four strategic imperatives which, we believe, will prove valid whatever shape the future takes.

The first is that we must pay intense and scrupulous attention to customer needs and go with them wherever they wish to go. They, above all, will shape the future and we can only hope to be part of that future to the extent that we help and support them. And as the telecommunications needs of our customers become inexorably global, it produces a serious dichotomy for those companies who aim to provide excellent service and value for money. On the one hand, excellence will be defined by responsiveness and customization of services, or a polymorphic approach. In contrast, value for money depends upon efficient resource allocation, globally – which is very much an isomorphic approach. Those companies who achieve the best trade-off between these forces will be the most successful.

The second principle is that we must recognize and accept the fact that meeting customers' needs requires the industry to work in new ways. Old rivalries and old demarcations will not be in the customer's best interest. In a sense we are lucky that the world's national telecommunications players learnt long ago how to co-operate, so that they could provide seamless, end to end, services for their mutual and separate customers.

The trick we are having to pull off now in the newly liberalized markets is to learn to compete (where that will benefit our customers) without un-learning how to co-operate (when that is what will deliver to our customers what they want and need).

Third, we must radically rethink our products and services portfolio. The 1980s saw a massive increase in quality and functionality in bespoke, private systems. In the coming ten years similar levels of service will roll out into the public networks, supported by the provision of network intelligence – by improvements in access technology – and by ISDN. All of these technologies are on our drawing boards now, and in many cases worldwide standards have been agreed. Once they are in the public domain, the days of true private networks will be numbered. Communications in the next century will be characterized by an explosion of new services offered on the public networks, with basic transmission capability becoming a commodity service.

With that in mind, we in BT have progressively shifted the focus of our R and D capability away from basic electronics at the component level towards software and systems engineering. It is in this area that we shall achieve differentiation over our competition. We have built a formidable capability in software engineering, devoted to network management and the development of new network services and systems infrastructure – and we intend to use it.

Fourth and finally, we need to rethink, quite fundamentally, the way we are organized and the way we work. Much of the discussion about the information revolution is really a discussion about responsiveness: how fast can we respond to customer needs? We can only deliver the level of customer service that will be required in the future if we have our own house in order. If our internal systems are not excellent we will be left behind. Products may well be simple in the future, but they will be offered by organizations who have developed immensely flexible – and complex – processes. And that means getting the balance of R and D right.

If these are the imperatives on which BT's strategy is built, what is the strategy? Well, the demand for – and range of – telecommunications services grew dramatically throughout the 1980s (growth in call volume consistently outstripped growth in GDP) and we believe that telecommunications will continue to be a growth area in the 1990s and into the next century. The challenge for us is to ensure that we are able to benefit from this growth, against a background of increasing competition.

BT's strategy is simple and, even more important, focused on well-defined objectives; it concentrates on the things we know how to do well – the running of telecommunications networks and the provision of services across them, both in the UK and internationally. There are two prongs to our strategy: the domestic and the global.

Taking the UK first. The vast majority of BT's revenues and profits (now and probably beyond 2000) derive from activities in the UK. Our UK strategy reflects the increasing levels of competition we are facing and is about our response to the erosion of market share. It has a defensive element, but is not a defensive strategy. It is about creating the right balance between holding what we have and seeking opportunities for new growth by paying meticulous attention to customers of all kinds, not just the supposedly more glamorous ones.

We are most vulnerable in our dealings with those customers who, as I said, will play a major part in shaping the future: intensive users of telecommunications services, both individuals and companies. We will fight to retain our position with these customers and to win back those we have lost; and we will do this through the quality of our services and the innovativeness of our service and pricing packages. The real cost of telephone services has fallen by some 25% since 1984. And we have achieved major progress too on quality of service. Nowadays, fewer than 1 in 200 calls fails due to network problems and a customer is only likely to experience a line fault once every 5 years.These trends need to be consolidated and reinforced.

If we are to deliver the services that customers want at prices they find attractive, we have to reap the productivity improvements that new technology makes available. Our investment has been substantial: more than £13 billion has been spent on expanding and modernizing our networks and services in the past five years. Over the next decade we plan to offer the benefits of digital technology to all our customers. IT facilities, (which only a few years ago were the jealously guarded prerogative of major corporations, defence establishments and the better funded universities) will be generally available, and they will be available to our domestic and small business customers as well as to the big spenders.

So that's the strategy for the UK. Turning now to our international strategy, the majority of profits currently derive from our activities in the UK. But even if we were only making, say, 10% from our international activities by the year 2000, that's still an awful lot of money. And it needs an awful lot of attention.

Both the structure of the international telecommunication industry and the requirements of our major customers have changed radically in recent years. In the early 1980s, most countries had their own national network operator (usually a state-owned monopoly) and national equipment suppliers, co-existing in a national and protected market. With liberalization and de-regulation this state of affairs is crumbling away, and the future belongs to those who can think and act globally and enter into strategic partnerships with others.

In addition, the customer base of the telecommunications industry is itself going global. Multinationals will be increasingly eager to avail themselves of the services of a truly global supplier – one who can design and manage global networks on a one-stop shop basis. We simply cannot afford not to provide such services; remember, many of the biggest multinationals are UK-based and others have UK-based subsidiaries; if we cannot serve their global needs, by the same token we will lose their business here in the UK.

Although in the next few years our main international business will be the traditional one of switched voice and international private leased circuits, we intend to develop a broader capacity, by, for example, building up our managed data network business, whose origins lie in our acquisition of Tymnet in 1989. We will not, though, make acquisitions for their own sake; once again, the emphasis is on focus and keeping it simple.

In his autobiography, *The Summing Up*, Somerset Maugham argued from a lifetime's experience that you should do only those things which only you can do (provided, and it's a big 'provided', that you can afford to pay someone to do the other things for you). In his view, a writer should be writing, not papering the hall or doing his or her accounts. It's all a matter of energy efficiency.

A similar precept applies in corporate life. Companies are best off doing those things which they do best. In strategic terms, knowing where you don't want to be and what you don't want to do is as vital as knowing where you do want to be and what you do want to do. BT is not in manufacturing, nor is it in financial services, or the entertainment business. We will not be beguiled by stand-alone domestic opportunities overseas, such as local privatizations or adventures in Eastern Europe, nor will we be seeking minority mobile stakes in other countries, unless they serve our broader plans for a presence there.

What this means is that BT has by far the most focused strategy of any of the would-be global players in our industry. And our customers appreciate that.

As predicted above, the future will expose the irrelevance of geographical and organizational constraints, which only recently we took for immutable facts of life. I also suggested it would require us to rethink the meaning of competition and necessitate creative collaborations, joint ventures and so on. In this context, I hope I will be forgiven for saying a little about the Syncordia venture which BT launched last year and which is very close to our hearts. Syncordia is a new company headquartered in Atlanta, which will provide voice, data and image services over an international network with end-to-end systems management and customer service.

We have launched the new company now because it is clear that the major multinational customers are struggling under the burden of managing highly complex and costly networking operations, which may be of major strategic significance but are no part of their core business. Syncordia will relieve them of that burden – by offering them a single source of help for managing all or part of their network and providing them with a seamless set of communications services worldwide.

Syncordia is, at one and the same time, evidence of the focused nature of our strategic thinking and a symbol of our commitment to the future. Imitation is the sincerest form of flattery and I have been pleased to see how seriously our global competitors have taken this initiative by BT.

I began by suggesting that it would be a mistake to try and picture the future in too much detail. However, perhaps even more counterproductive than attempting to base a strategy on a fixed and detailed picture of the distant future, is to assume that the future will be like the past, and that all we need to do is do what we have done before, only better and more of it.

Blithe futurology may lead you down blind alleys; but assuming that the future will look like the past with knobs on results in organizational sclerosis, an irreversible and fatal hardening of the arteries.

We have to offer customers not futuristic technological solutions, but goods and services that will transform their lives today and tomorrow. And then, when they change their minds, want improvements, want something else, we have to go with them. We have to recognize that we have a range of responsibilities, not just to customers, but to the people who work for the company, shareholders, governments and to Society with a capital 'S'. We have to achieve that balance in such a way that truly adds value to the world in which we operate.

No other strategy for the year 2000 and beyond has a chance. And if you need any further convincing, remember that George Orwell's 1984 has come and gone, that Mrs Thatcher is on record as saying she didn't expect to see a woman Prime Minister in her lifetime, and that the BBC weatherman promised us that there wouldn't be a hurricane.

So can we really feel comfortable that in the Third Millennium we will not all be put out of business by telepathy?

2

The American scene

J.S. Mayo, President, AT&T Bell Laboratories

This article will discuss the infrastructure that delivers the benefits of information technology to society – benefits in the form of new communications products and services. It will also examine some changes that are needed in that infrastructure to meet the societal needs of an increasingly global future. Then it will examine some of the major impacts that information technology could have on society, and how the marketplace could affect the evolution of those impacts.

As we explore communications after the year AD2000, the 'American scene' must be examined in the broadest way possible – in the context of powerful interrelated thrusts in information technology, in competition, and in globalization. The telecommunications infrastructure in the USA today is the beneficiary of the powerful twin forces of information technology and competition. The third force, globalization, is both a product of telecommunications and a driver of telecommunications. And these forces are linked inseparably and directed by the goal of satisfying the information needs of end users – needs that are increasingly on a global scale.

In America, we believe the key issue is not *where* we are headed in communications from the user's viewpoint, or what technology will take us there, but *how* the task will be accomplished.

Let me briefly review the issues of where we are headed and what technology will help us achieve this goal. Beyond the year 2000, the forces of information technology, competition and globalization will be instrumental in achieving the long-standing goal of the telecommunications industry: the goal of providing voice, data and images in any combination, anywhere, at any time with convenience and economy. This is very much a global goal, dictated by the increasingly global needs of end users. Moreover, we will see spontaneously provided information, communications and entertainment – and the merging of all three into

Communications After AD2000. Edited by D.E.N. Davies, C. Hilsum and A.W. Rudge.
Published in 1993 by Chapman & Hall, London, for The Royal Society. ISBN 0 412 49550 3

multimedia communications. These capabilities will be made possible by a highly intelligent, high-capacity multimedia network – one that will be accessed by a multitude of advanced multifunction terminals. And the various types of information terminals on the desks of people will be gateways to the intelligence stored in switched networks around the world.

Key information technologies such as microelectronics, computing and photonics are helping to make all this possible. And these technologies are expanding exponentially in capabilities. For example, the number of components on a silicon chip – one measure of growing microelectronic power – continues to double roughly every 18 months. The processing power of microcomputer systems continues to double every year. And the overall capability of lightwave transmission systems continues to double every year. These trends have great momentum because they are supported by solid science and technology bases, by enormous capital and human resources, and by powerful customer demand. Such trends enable us to project what technological capabilities we will have beyond the year 2010. The key information technology of software, which is the glue that holds the other technologies together and makes them work in systems, has traditionally been a bottleneck to progress. But even software is yielding to increases in programmer productivity from growing reuse of previously developed and tested software modules. In addition, the key information technologies are enabling rapid progress in emerging technologies such as speech synthesis and recognition, and video, which also are helping to achieve the broad telecommunications goal.

That, then, is the goal of the US telecommunications industry, and the information technologies that will help achieve it. As noted, however, the key issue in the future of US communications is *how* the task will be accomplished. And that involves the way we handle the merging of entertainment with computers and communications – and, above all, how we reshape the infrastructure that supports and delivers information technology and systems to meet society's accelerating demands for new products and services.

Let's look broadly at this complex infrastructure. In part, it includes political bodies at both the state and national levels, most notably the US Congress. It also includes regulatory bodies at both the state and national levels, most notably the Federal Communications Commission or FCC. These political and regulatory bodies impact a marketplace system that is vital to matching information technology solutions to the needs of end users. The marketplace provides the familiar 'pull' to complement the traditional 'push' of technology, both of which combine to help meet user needs. In addition, the infrastructure includes standards bodies at the national and, importantly, at the regional and international levels –

bodies that range from the Institute of Electrical and Electronics Engineers (IEEE) to the International Telecommunications Union and the International Organization for Standardization. There's also the growth of new industry consortia aimed at facilitating the marketplace introduction of products and services that comply with new standards. The infrastructure also includes a private-sector research and development system – of which AT&T Bell Laboratories is a part – that generates new technology and converts it into new products and services.

We must reshape this infrastructure in order to produce more timely political and regulatory decisions and standards, and ever-shorter intervals for converting technology into new products and services. This is a broad challenge, and clearly one whose scope extends beyond the US – because of the strong influence of globalization. Let's examine the status of this reshaping in a bit more detail.

Consider the political and regulatory arena. Even though they are separate, the two will be discussed together – especially since their thrusts are often complementary.

Until about ten years ago, global communications grew on a diet of pure technology. But in the 1980s, competition was added to the mix. In the US we trace the genesis of that competition to the breakup of the Bell System. (For this singular event, I should add the influence of the US judiciary system to that of the political and regulatory arena.)

The Bell System – a highly centralized, regulated monopoly for most of a century – was thoroughly splintered into eight companies related by technology and a heritage of service, but otherwise with very different goals. Under the terms of the 1982 Consent Decree agreement, AT&T was required to divest the local parts of the 22 Bell operating companies – the parts providing local exchange or local access services to long distance and international networks. Ownership of customer premises equipment was to be retained by AT&T along with ownership of Western Electric, Bell Laboratories and the AT&T Long Lines Department. In effect, the decree separated the monopoly parts of the Bell System's business from the competitive parts. And AT&T was permitted to enter the commercial computer market and other arenas beyond the regulated telecommunications business.

Today, except for local service, which is still a monopoly bottleneck, competition is rampant in every facet of the US market – to the point where foreign companies enter easily. Choices proliferate among capabilities and innovative providers. Prices are lower, and products and services offer far more features than those common ten years ago.

Even after ten long years, however, AT&T remains the only regulated long distance carrier, as US government oversight has failed to keep pace with the shift toward an open marketplace. Despite this, today the US telecommunications industry is stronger, more vibrant, and more attuned to customers than ever before. AT&T has become a stronger company. The marketplace has forced us to change, and the change has been for the better.

Even so, America's handling of deregulation and competition is not necessarily a perfect model for the rest of the world. But even an imperfect model can be useful. As we at AT&T look around the globe, tightly controlled markets are still the rule, as many governments still protect native telecommunications providers in their home markets – many of the same providers who freely enter the US market.

One can note with optimism, however, that changes comparable to those made in the US telecommunications industry have begun in Japan and the United Kingdom, and are occurring in nations throughout the world. Particularly notable have been the European Commission's attempts to unify and liberalize Europe's diverse and fragmented telecommunications market – an effort that must succeed if the European Community is to fully achieve its enormous economic potential.

The national, regional and international standards-setting process is another vital element of the infrastructure that delivers information technology to meet societal demands for new products and services. The broad goal of standards-setting in my industry is connectivity, compatibility and open networking of communications and computer systems from multiple vendors – whether in common carrier public networks, in private networks or on users' premises. AT&T fully supports standards bodies working toward this goal.

There are several types of standards, with differing levels of priority. Clearly, the most vital and broadest standards are the architectural standards. Examples range from the so-called Integrated Services Digital Network or ISDN interfaces to the Open Systems Interconnection Reference Model. Such standards affect the architecture of systems and services and are vital enablers, benefiting all players. Moreover, these standards should have first priority because they are needed first, before implementing a technology. Failure to do so erects the most serious potential barriers to the progress of interoperability. The obvious danger is a growing embedded base of nonstandard products, perhaps with significant incompatibilities from one vendor to another. This is largely the situation today, for example, in the world of data processing.

The second highest priority is the technology-driven standards – those for which the technology typically precedes the standard. The voluntary consensus standards-setting process now used is particularly important for these standards. Attempts to set such standards without consensus can easily run the risk of setting standards too early, before the technology has matured. So the standards might not only be inappropriate, but they also might diverge from international directions, weakening the global standards-setting process and the competitiveness of nations.

The third-priority standards are those for which basic technology is not a deciding factor – for example, standards that relate to the encoding of generic service or maintenance capabilities. Because entire technologies are not at stake, such standards can be somewhat arbitrary, in the sense that any number of viable solutions are often available. And since the matter *is* arbitrary, it can and should be settled quickly.

The lowest-priority standards are those that we really do not need at all. For example, standardizing the innards of systems or many incompatible variants of the same thing or defining every potential feature or capability can be a drag on standards-setting resources and an irritant to both vendors and consumers.

Such efforts can further lengthen a process that, by consensus, already takes too long. My overall point is that we must focus our energies where they will do the most good. Above all, we must resist efforts to politicize the standards-setting process and to bend it to the competitive advantage of individual players. With the growing trend toward globalization of telecommunications, the international standards-setting process and the development of more timely standards will become increasingly critical.

Let's look now at another vital element of the infrastructure that generates new technology and converts it into new products and services. Let's look at the private-sector research and development system, and especially at the R&D system within the communications industry – the system that's made up of organizations such as AT&T Bell Laboratories. Perhaps no other aspect of this system is more critical to accelerating the conversion of technology into new products and services and therefore to international competitiveness than an unwavering focus on the needs of customers, both domestic and international. I will draw upon the example most familiar to me – that of Bell Laboratories.

AT&T strives to enhance its customer focus through organizational structure and through the pervasive application of quality principles.

As you may know, AT&T is organized into a number of business units, each focused on particular types of customers. We at Bell Laboratories, in turn, support those business units with the technologies they need now and into the future. In fact, the Bell Labs development organizations are aligned or partnered with the business units they support. That structural arrangement enhances both our understanding of customer needs and our responsiveness to those needs.

The pervasive application of quality principles complements this approach to customer needs, and is instrumental in our striving for customer satisfaction. That's because we have learned to define quality in terms of customer satisfaction. The traditional supplier's definition of quality focused heavily on the product or service itself and its conformance to drawings and specifications. This traditional concept of quality usually focused, for example, on issues of form and fit, and on whether the product worked. Too often, what the actual customer thought was immaterial.

With today's quality revolution, we all speak of quality that is a strategic differentiator for products and services, of quality that is also a cost and time saver, and of quality that – above all – is the key to customer satisfaction. Because of its vital link to customer satisfaction, quality must ultimately be defined by the customer, both domestic and global. So, across our industry, today's definition includes all that we traditionally associated with quality – plus much more; plus features the customer wants; plus timeliness the customer wants; plus a price the customer is willing to pay.

These organizational and quality thrusts drive Bell Labs' alignment with both customer needs and business needs. These alignments extend through development – and, for an entirely new product or service – all the way back to research or exploratory development. And the alignments are vital to the absolutely critical process of technology conversion, of moving technology to the marketplace in the form of new products and services. Technology sitting on a shelf or in a research lab is virtually useless. In the arena of global competition, the key is how *fast* technology can be moved to the marketplace. Accelerating technology conversion often involves concurrent development and the reengineering of development processes in order to shorten development intervals. And such ever-shortening intervals should ultimately lead to overnight customizations.

These, then, are some aspects of the need to reshape the infrastructure that delivers information technology to meet society's accelerating demands for new products and services. Let's now examine some of the major impacts that information technology could have on society, and how the marketplace could affect the evolution of those impacts. At the very least, information technology should have a broad impact in reshaping and expanding communications and entertainment markets, and on creating entirely new markets.

It's tempting to view a future based on the projections of technological trends alone. Unfortunately, such trend data does not include willingness to pay. As I noted earlier, marketplace 'pull' is a vital complement to the 'push' of technology – since technical feasibility alone cannot bring about a technology-based vision of the future. Never before, in fact, has the interaction between technological forces and societal systems been so critical to the shape of the communications infrastructure.

Telecommunications will be shaped, in large part, by social needs and desires yet to be uncovered. For example, when he was a US Senator, Vice-President Albert Gore introduced legislation that would create a gigabit national data network at the turn of the century. Although there is talk of high-speed distributed data networking and networked supercomputing, no one at the moment has a clear conception of what such a network will be used for. No one doubts, however, that the network will be built and that a range of uses will materialize.

With these thoughts in mind, let's glance at the potential impacts of ISDN or Integrated Services Digital Network, wireless communications, speech synthesis and recognition, and video.

ISDN is an international standard, in part, for combining voice, data and images on the same communications line. Over the next two decades, it will mature and will change the way we work and play. In fact, ISDN has probably been slow to catch on today, in part, because – like the telephone – it requires us to change the way we do things. Plain Old Telephone Service or POTS, as we call it, may be largely replaced by ISDN as the basic wired service, as ISDN becomes ubiquitous. High-fidelity voice and high-quality videophone may be two of the many benefits of ISDN.

Data bases – for both home and office – need to be tied to ISDN to enhance the efficiency of our home and work lives. ISDN should begin in earnest the work-at-home movement that has been discussed for so many years, but whose easy feasibility has eluded us. Already, tests with several types of jobs show ISDN alone can make work at home more efficient than work at the office. And telecommuting workers will no longer have to worry about bad weather, crowded highways, or living close to work. In the US, for example, we are seeing a steady, alarming growth in automobile traffic without a concomitant growth in highway capacity. This situation cries out for the displacement of this traffic through telecommunications – and ISDN may be one answer. Because work at home changes family life, however, cultural evolution – in addition to technical feasibility – is required there before the concept can flourish to its full potential. And that potential, incidentally, could include school at home and self-paced learning. What a societal change that could create!

The strong trend toward wireless networking could bring an equally broad societal impact. The microelectronics technology I mentioned earlier makes possible increasingly intelligent communications networks, and these in turn enable the wireless trend.

Because people are increasingly on-the-move, a very large number of network end links will be wireless. Freed from their wire tethers to networks, customers will have great mobility, and will depend on Personal Communications Networks or PCNs to automatically direct calls to them wherever they are. Just as microcomputers brought computing to people, so will PCNs and wireless technology bring improved communications to people on the move. In fact, the ubiquitous voice service may move to PCNs. PCN users will be tracked via intelligent network data bases, which will follow them as they move from one micro-service area to another. And people will need personal telephone numbers that will follow them wherever they go, independent of location, so that they can be reached anywhere.

This personal telephone number is potentially the most sensitive aspect of the new PCN – because there is a fundamental trade-off between access availability and privacy. How will people determine who shall be permitted to call them? And how will they be able to screen callers for this purpose of access control? This will surely be a sensitive social issue.

In addition, data communications also will travel over PCNs. There will likely be dramatic growth in the use of wireless notebook computers. And the equally likely use of pen-based input will make these computers easy to use for both business and personal applications. Moreover, the US growth in automobiles mentioned earlier also argues for more mobile services based on wireless communications – such as navigation, entertainment and information services in automobiles.

Another information-technology area of great potential social impact is speech synthesis and recognition. Much progress in this area is being driven by the advancing power of microelectronic chips. Automatic speech recognition and synthesis will become commonplace, and will result in a variety of service capabilities based on the ability of intelligent machines to talk and listen much as people do.

Within 20 years, we should have customized speech translators. Speech in one language would be automatically translated into a second language, which might then be synthesized with the voice characteristics of the original speaker.

Moreover, speech recognition vocabulary will be unrestricted, and will permit natural language interaction. Humans serving as operators and order takers will graduate to more interesting jobs as this capability begins to permit direct interaction with communications networks and databases. In fact, most business phones could be answered by machines, not by humans. But that applications transition would take many decades, even after the technology is capable. Although this trend could dramatically increase business productivity, there is the major question of human acceptance of this new impersonal approach.

The last information technology with broad societal impact that I want to consider is video. Projections for the US video market show the appetite for entertainment is enormous. That appetite will become a powerful marketplace incentive for the evolution of an all-fibre communications network, including fibre to the end user.

Although there are few videophones today, it has been estimated that in 20 years there will likely be 20 million lines in the US – accompanied by on-going debate over user privacy. (AT&T, of course, introduced the first colour videophone that operates over conventional copper lines.) And 20 years hence there may also be 60 million High-Definition TV sets in the US alone. The appetite for video and HDTV is sure to increase as applications expand.

There are several trends that suggest entertainment video will eventually become both interactive and available on demand. By its nature, this type of entertainment tends to be private. Today, people gather as groups to watch TV much less often than they did in the early days of TV. The availability of entertainment video is increasingly reflecting this desire for TV that is basically personalized to the needs of individuals. For example, consumers will be able to enjoy customized showings – at any desired time of the day or night – of movies on a list of box-office hits. And some movies and live programmes may be designed to be interactive with the viewer. The potential impact on the entertainment industry is profound.

All this, along with the technology trends, suggests that 'virtual reality' will be a hot topic in about 20 years, building on the capability for widespread interactive telecommunications and entertainment. This will enable people to indirectly and remotely experience a place or an event in all dimensions. This capability, for example, will allow people to attend meetings without travel, although such telepresence will not displace all travel.

Virtual reality also will enable a person to 'sample' a new home or a vacation destination; and to select a school, a car, or virtually any other purchase that now requires examination in person – all without ever leaving home. The individual will be able to turn his or her head, for instance, and see different aspects of a particular scene – or be able to feel the texture of wallpaper from afar, by using an electronic glove. Eventually, we will even smell distant flowers through electronic sensors. Virtual reality also will greatly enhance the concept of telepresence. We will be able to visit from afar – and do it much more frequently than we could travel. Using telepresence, we will increasingly work at home – and do so long before virtual reality has matured.

Needless to say, the evolution of virtual reality will depend heavily on human acceptance, on the marketplace. I suspect that there will be those who will decry the potentially isolating impact of the advance – those who will swear the advance transforms their homes into electronic and photonic prisons. Many others will hail the liberating flexibility of the new technology.

These are just a few of the societal impacts to be anticipated from information technology. The associated marketplace opportunities are abundant and often revolutionary. Although my topic has been focused on the American scene in communications of the future, the vision sketched is global in scope – driven both by the rich and powerful technology and by the increasingly global needs of end users.

No single company, regardless of its size or technical sophistication, will alone be able to realize this vision and these opportunities in meeting the needs of ever-greater numbers of global customers. What certainly will be required is partnerships among the global telecommunications providers. And partnerships required to deliver the benefits of information technology may be as complex as the technology itself. Such partnerships will work best in countries and markets where there is reasonable freedom to compete. Indeed, AT&T, British Telecom and France Telecom have already teamed up to begin creating a global network for at least one large global company. Furthermore, the concept of scale tells us that the world is becoming so small and the cost of information technology so great that there can be far fewer suppliers of the technology than there are countries in the world. Such factors of scale will decrease the numbers of information-technology suppliers for the future, and will reinforce the need for partnerships among those suppliers.

The broad service opportunities outlined here are just a small step toward the long-term goal of the US telecommunications industry: to provide voice, data and images in any combination, anywhere, at any time with convenience and economy.

This exciting vision will involve the sophisticated needs of ever-larger numbers of global customers, and will require open markets and free competition, universally agreed-upon standards, and international co-operation and partnerships of a scope never before attempted.

This vision cannot be realized for the US alone, for the 'American scene'; it is a global vision, and the US telecommunications industry stands ready to carry out our role with our global partners.

3

Future technological developments and their impact

G. Tenzer, Member of the Management Board of Deutsche Bundespost Telekom

Telecommunications will be determined in the future by such key parameters as: technological advancements in microelectronics and photonics and the resultant great diversity of telecommunications services; the rate of economic growth and thus the demand for these services; and last, but not least, the regulatory environment and thus the freedom of action accorded to network operators. This article proceeds on the assumption that the last two factors are stable and will therefore focus on future technological developments and their impact.

Over the past decade, telecommunications and information processing have developed at a rapid pace. Different forms of communication, such as the telephone, fax, data transmission, and value-added services all offer an enormously increased range of services and performance features. The rapid development of these technologies in recent years has been accompanied by far-reaching structural changes. Telecommunications and computer science, which used to be relatively independent, will grow together thanks to the advance of software-supported system solutions. The merging of telecommunications and computer science into 'telematics' will provide the economy with enormous synergies: the traditional factors of production, namely capital, land and labour, can thus be virtually supplemented by a fourth one – telematics.

Telematics makes it possible, on the one hand, to rationalize internal and external corporate workflows and thus considerably increase productivity. On the other, it enables products and services to be improved. The huge range of process and product innovations supported by new telecommunications technologies makes it possible for enterprises, i.e., for our customers, to adapt flexibly to specific and rapidly changing

Communications After AD2000. Edited by D.E.N. Davies, C. Hilsum and A.W. Rudge. Published in 1993 by Chapman & Hall, London, for The Royal Society. ISBN 0 412 49550 3

markets. Telematics thus has a vital influence on key economic factors such as employment, economic growth and competition and is, in the end, the basis which guarantees our standard of living. Within a few years it will most probably be the most important sector, or one of the most important, sectors of the economy.

A carrier like German Telekom, the largest telecommunications operator in Europe and the third largest in the world with an expected turnover of more than 50 billion DM and record investments amounting to 30 billion DM in 1992, has a special responsibility in this expansive market.

It is therefore our objective to turn the enormous opportunities derived from this technological change into optimum, tailor-made applications in order to meet present and future customer requirements in an even better and more economical way. These technological changes are documented in the following scenario.

We are at present moving from analogue to digital technology in all areas of our telecommunications network (transmission, switching, and terminal equipment). We are going from copper to optical fibre systems, from static to mobile communications and from hardware-oriented to software-oriented technologies. The most important innovations over the past few years were:

- the rapid, continuous progress made in microelectronics and the required software technologies – the 'technological locomotive memory chip' has increased its integration density four-fold every three years and has thus become a central telecommunications component making it possible to digitize switching and transmission systems economically; and
- similar developments in the field of optical fibre-based communications and associated systems, which, with their virtually unlimited bandwidth, are by far superior to the classical copper cable. In this field too, further components are being developed such as optoelectronic ICs, optical repeaters and switching matrixes, which will considerably improve the efficiency and economic effectiveness of optical fibre transmission.

The interaction between these innovative forces as well as the substantial cost reductions connected with technological progress are pushing the development of communications networks towards:

- Greater flexibility in providing links at short notice;
- Higher intelligence, making it possible to offer the customer a wider range of services and service attributes through central databases;
- Greater transparency and higher bandwidths, where the transmission

path can be used by a larger variety of telecommunications services; and

- Increased internationalization, whereby the introduction of European and international network infrastructures take the global business activities of customers into account.

On the other hand, these new technical possibilities have led to a considerable change in demand for telecommunications services. Whereas customers had to manage with a few standardized services in the past, users now request customized solutions with a variety of modern service attributes and a high degree of mobility.

Interaction between process and product innovations and the resulting synergies will quickly improve network efficiency while leading to cost reductions at the same time.

What is DBP Telekom's entrepreneurial response to these fundamental changes?

Telekom's concept for the future is determined by two essential factors.

1. Investments in our networks account for 60% of Telekom's costs. This means that we should buy the most favourably priced systems on the world market.
2. The second factor takes us back to the subject of customer orientation. We must use the technology that offers our customers those features that are demanded on the market. Given the long break-even times of

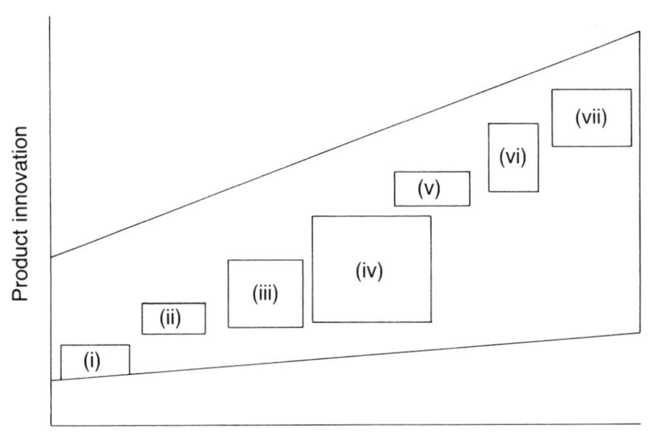

Figure 3.1 Evolution of telecommunications: (i) analogue network, (ii) digital network, (iii) integrated services digital network (ISDN), (iv) Synchronous Digital Hierarchy (cross connect NKÜ 2000, REBELL), (v) intelligent networks, (vi) fibre to the home, (vii) broadband ISDN; ATM.

telecom networks, we must select systems that are future-oriented and upwardly compatible.

Our strategy over the next few years will therefore be characterized by the following cornerstones:

- swift continuation of the digitization of switching equipment and further development of the system towards an 'Intelligent Network', which makes it possible to provide new services and service attributes in an ISDN network;
- continuation of the digitization of the transmission equipment and introduction of the SDH technology, incorporating modern network management;
- expediting the most effective use of optical fibre technology in the loop for the existing telecom services; and
- establishing an optical fibre network for individual broadband communication, incorporating ATM exchanges and transmission techniques.

These activities ensure that the customer will be gradually offered a growing range of user-friendly services and more flexibility in the service attributes provided. In addition, the optimization of the use of all transmission resources will lead to substantial cost reductions.

This development is a continuous process of technological innovation, where new, more efficient generations of equipment are initially added to the existing ones – where the need is greatest – and then gradually replace them in the subsequent phase of penetration as costs are steadily reduced and market acceptance grows. As a result of this dynamic process, several generations will always have to be operated simultaneously. This means that our networks permanently have an overlay structure, coupled with the ensuing adaptation problems. Therefore, the transition process must be optimized with suitable launching strategies to ensure that the benefits of productivity can quickly be reaped. At the same time, it will be a matter of promoting and increasing market acceptance through product innovation and diversification, supported by an attractive pricing policy.

In view of the heavy investments and the business risks involved in the long-term nature of such investments, Telekom will only be able to undergo this development in an evolutionary manner. This means that we have to consider not only our mandate in terms of the national economy, but also our business goals and the general conditions prevailing.

Telekom has made substantial investments over the past few years to digitize the switching and transmission network. Our plans to digitize switching centres envisage digitizing the trunk network nationwide by 1993. Complete digitization is to be achieved by 1995.

At the upper network level, 80% of the transmission network has been digitized. The standard utilization of optical fibre technology in the trunk network will thus provide the basis for high-channel transmission systems with at present up to 2.5 Gbit s^{-1}. A massive increase in transport capacity is a precondition for cost reductions on a channel-per-km basis.

In the next few years this will also be supported by 10 Gbit s^{-1} systems currently being developed. If the trend towards lower transmission costs is to continue in the future, we must ask ourselves whether we can afford to concentrate on the development of increasingly powerful codecs that enable us to achieve only less than optimal results rather than installing more powerful and cost-effective networks which would be more favourable for customers and therefore promise to be more successful.

While the trend towards increased traffic concentration leads to declining average costs per channel km, these economies of scale can only be realized if a sufficient occupancy level can be achieved by having a high degree of internal flexibility when switching transmission flows. Besides, customers – commercial customers in particular – demand that circuits and individual applications be provided swiftly and at reasonable prices.

The plesiochronous multiplexing equipment with mechanical distributors which has been used so far does not meet these requirements to a satisfactory degree.

Implementation of the new synchronous digital transmission system SDH makes it possible for us to save costs in order to improve the economic efficiency of this network, to further decrease the demand for future investments and to improve at the same time also the service attributes offered to our customers.

In contrast to the former plesiochronous systems, where the transmission band had to be demultiplexed in several stages to get the desired channel, we can now access the transmission channel needed with the help of new and synchronous systems. This process innovation is as important as the transition from manual to automatic switching systems. It is one of the outstanding milestones in efforts to design future networks.

With our new network node NKÜ 2000, which will allow unrestricted and remote controlled switching of individual channels and also replace today's mechanical frames, we will for the first time have a technological system which allows flexible activation of transmission capacity and a sufficient occupancy level of the transmission systems.

In the future, we will thus be able to offer users of transmission paths various bit rates (Figure 3.1) ranging from 64 Kbit s^{-1} to 2 Mbit s^{-1} and 34 Mbit s^{-1} up to 140 Mbit s^{-1} at lower prices. Time required to provide such transmission paths will drastically decrease to only minutes, and

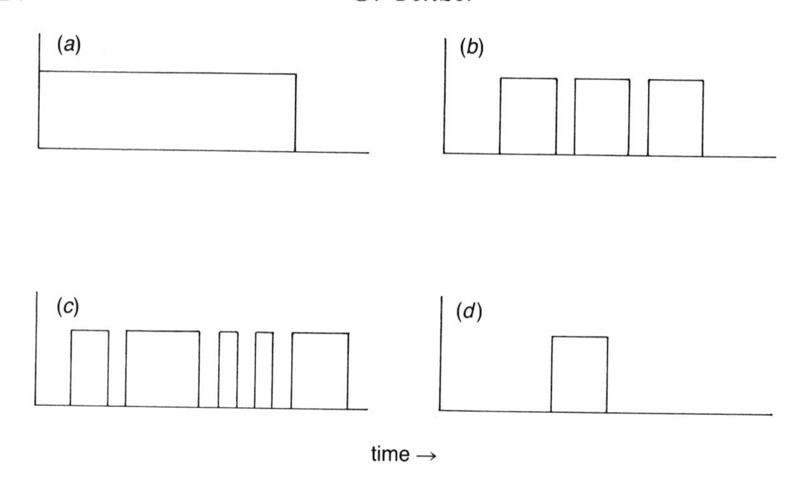

Figure 3.2 Activation of circuits in the synchronous network: (a) permanent, (b) cyclical, (c) sporadic, (d) non-recurrent. Bit rates: 64 Kbit s^{-1}, 2 Mbit s^{-1}, 34 Mbit s^{-1}, 140 Mbit s^{-1}. The activation time is a few seconds or minutes, and the activation is either self-activation or DBP Telekom.

users will be free to choose any time required. It will also be possible to set up so called 'shadow networks', i.e., complete network configurations which can be activated on demand from a central station and at short notice.

A first synchronous partial network using 17 NKÜ 2000 and synchronous line equipment with bit rates of 622 Mbit s^{-1} to 2.5 Gbit s^{-1} will be installed in northern Germany this year.

So far, network structures have been determined by the decadian network architecture, the limited capacity of switching equipment, rigidly defined transmission resources and multiplex equipment etc. All these individual parameters have changed substantially due to the new systems. The high efficiency of digital switching centres in combination with common channel signalling and an enormous increase in capacity and intelligence in the transport network allow us to move away from the fixed and static network towards a flexible network structure with fewer switching nodes. Network layers and network nodes can thus be decreased, which again helps to reduce considerably investment and operational costs.

Another step to be taken within process innovation is the interconnection of synchronous systems with an efficient network management system. This interconnection is a precondition for the computer-based control and monitoring of the entire transmission network from one central workstation. The following possibilities are to be offered with one

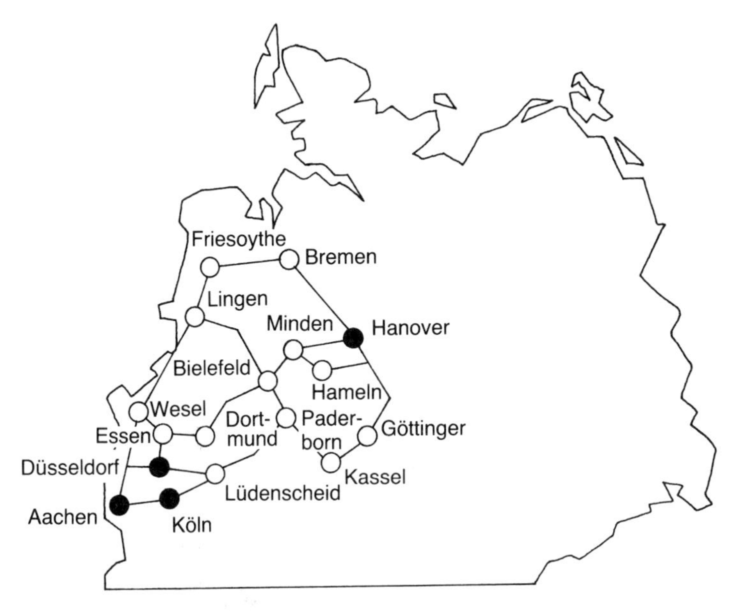

Figure 3.3 Initial size of the synchronous network in 1992; (●), VISYON pilot trials; (○), NKÜ 2000 locations.

such network management system, termed the 'Telecommunication Management Networks', the components of which are currently being standardized internationally.

- automatic fault control
- automatic network configuration
- protection against unauthorized access
- automatic evaluation of network data used as a basis for billing, statistical evaluations on network quality.

It will only be possible to make full use of the entire potential of process and product innovation within digital telecommunications technology, if present analogue transmission systems are also completely replaced by digital equipment in the local network.

We will therefore replace analogue switching equipment with more efficient digital switching systems as soon as possible. This is an important precondition for implementing reasonably-priced service attributes which are already quite popular in the US, for example, and can account for up to 10% of overall turnover.

Another step to introduce new services in narrowband networks much faster and more flexibly on a nationwide basis is the concept of the 'intelligent network'. This concept involves separating the actual physical call switching and the service-specific 'intelligence'. The concept of the intelligent network thus makes it possible to arrange service attributes freely and in an entirely transparent manner without making nationwide and comprehensive pre-investments.

In a nationwide operational trial which will start in late 1992, four services, already available today, will be provided in the intelligent network to begin with: Service 130, Televotum of Teledialogue Teleinfo Service and Service 180. Later on, supplementary services requiring a higher degree of intelligence in the network will be implemented. These include in particular three features, namely: credit card calls, virtual private networks, and wide area centres.

The efficiency and economic effectiveness of local networks are to be improved by additional measures such as the introduction of transmission equipment on access lines. The so-called 'wireless access line', which makes it possible to connect a subscriber quickly to an exchange using mobile communications if there is no copper cable available, is at present only used in what was East Germany. We are also investigating how this wireless access line system can be used economically and strategically on subscriber access lines. It will be important, in this context, to link DAL systems with a number of the latest network developments in an intelligent way.

Another step towards complete digitization of the network is swift introduction of ISDN. On the basis of the new digitized local exchange system, we are digitizing a local network line to subscribers' premises and are thus introducing ISDN. In addition to the products already available, customers will be offered new, more sophisticated telecommunications services with transmission rates of up to 64 $\mathrm{Kbit\,s^{-1}}$ in a fully digitized network, in a flexible and economical manner. Nationwide coverage is to be achieved by the end of 1993 – i.e., within five years after ISDN was introduced.

The most important technological step towards the long-term objective of installing an optical fibre digital broadband network is, however, the early introduction of optical fibres in the local loop.

The large bandwidth with the resulting low transport cost for voice and data as well as the improved quality and availability of transmission are the main advantages of optical fibre systems. Optical fibre has thus completely replaced copper in the trunk and local junction line networks. The real potential of optical fibre systems, i.e., the enormous bandwidth, can only be used to the full, if the fibre is brought directly to the vicinity of the customer or to the customer's premises. When extending the optical fibre infrastructure on subscriber lines of the telephone network, we are still running into economic obstacles: up to now demand for interactive broadband services could not justify an expansion of the overlay network.

In order to penetrate the market for individual broadband communication swiftly in spite of these obstacles, Deutsche Bundespost Telekom was the first carrier to put into operation a digital Switched Broadband Network suitable for subscriber dialling in February 1989. Telekom's pioneer work not only provides technical innovations in the network, but is also aimed at developing and testing, under real-life conditions, broadband services for individual communication needs in line with market requirements at an early date, in close co-ordination with users.

Up to 1000 subscribers can be connected to Telekom's switched broadband network, which is still the only one of its kind in the world. It gives them an opportunity to experiment with new forms of broadband communication and gather know-how in this sector.

The significance of this network is illustrated, in particular, by the videoconferencing service that is currently being operated on a commercial basis via the switched broadband network by some 300 private videoconference subscribers. This also demonstrates the great interest users are showing in this network.

As regards the development of broadband applications and the related terminal equipment, we are meanwhile focusing our attention on several areas, for example: computer interworking, telepublishing, telemedicine and the delivery of radio and TV programmes, to name but a few.

The close co-operation between Deutsche Bundespost Telekom and industry, various research institutes and users in this project ensures that the results achieved in technology will be realistic and can also meet future requirements. As regards application development, this close co-operation will certainly lead to broadband services that will be successful on the market if they are designed in line with customers' needs.

With its switched broadband network that was put into operation in 1989, Telekom is already providing a minimum optical fibre infrastructure for new broadband services that runs to a point near subscribers' premises and which operates at a transmission rate of 140 $\mathrm{Mbit\,s^{-1}}$.

However, in spite of these efforts, we still have not achieved a breakthrough for fibre in the loop.

By no longer considering demand as a factor in providing broadband services, we are now looking for new technical solutions which will permit the cost-efficient utilization of optical fibre transmission systems in the local loop, solutions which are based on interactive narrowband services already available or on the various standards for transmitting sound and TV programmes. This is the basis on which we will also replace copper with optical fibre in the local networks in the coming years; this will help cut down on the, in operational terms, inefficient parallel existence of different technologies. However, with this new strategy we are also keeping open the option of upgrading the new optical fibre systems in due course to offer cost-efficient broadband communication services as well. We are thus also establishing a basis for the further development and acceptance of such services.

To test the different new concepts for optical fibre systems and to be able to make a choice in line with our requirements, so-called 'optical access line' pilot projects (OPAL) have been launched in seven different locations situated in regions which have a varying degree of development and different customer groups.

Another measure to speed up the introduction of optical fibre systems in the local loop is Telekom's decision to provide 1.2 million homes in the former East Germany with connections for telephone, data communication and radio and television by 1995 using sophisticated optical fibre systems. Owing to this decision, we are the first network operator in the world to leave the trial stage and to start implementing optical fibre in the local loop on a regular basis. Again Telekom is able to maintain its leading position worldwide in the implementation of such future-oriented technology.

Thus, several technologies are being used simultaneously in the local network: the conventional copper cable, the new optical fibre cable and, increasingly, radio systems, which also include PCN systems. The PCN is a new, forward-looking technology that goes beyond the 'mobile communication' classification. When one considers that PCNs can serve a total of 30 million subscribers in Germany alone, it becomes evident that this technology provides the key to the mass market for mobile communications. It is not yet clear at this stage, however, whether this technology will be a substitute for, or an addition to the terrestrial local network.

The implementation of optical fibre, as a broadband transmission medium in the local network, is accompanied by the introduction of synchronous digital hierarchy (SDH). Parallel to the expansion of the SDH-trunk network, ring structures and subscriber accesses with flexible multiplexes are being tested in a pilot project called VISYON in the Aachen, Cologne, Düsseldorf and Hanover local networks.

These developments in the local network naturally had an impact on the topological structure in this area. With these new systems it is possible to have larger local exchange areas, which means that the number of exchanges can be considerably reduced, which was also the case in the trunk network.

Universal intelligent broadband ISDN using optical fibres on the basis of ATM technology represents the highest level of process innovation that has been reached so far.

Broadband ISDN is designed to flexibly meet any transmission requirements ranging from 64 Kbit s^{-1} to more than 150 Mbit s^{-1}. The network will initially consist of network nodes with asynchronous cell-oriented switching based on the ATM principle.

ATM nodes and the SDH-based transport network will be capable of switching and transmitting connections with different transmission speeds simultaneously and will thus be more capable of adapting to the different needs of users. This results in productivity benefits for the network because – unlike circuit switching – fixed capacity transport channels are not required to be provided for the entire duration of a call, but only for the period of actual usage.

Bearing this in mind, B-ISDN based on ATM technology seems to be capable of integrating various networks and telecommunications services ranging from the telephone to data, office and multimedia communication and on to videotelephony in TV quality standards. We are aiming at introducing B-ISDN in the mid 1990s. Here, we attach great importance to co-ordinating our approach internationally.

All developments in telecommunications clearly show that we are already on the path leading away from microelectronics to photonics. These developments are being implemented from the top down to the bottom network level, that is to say, down to end users. Additional steps will cover implementation of optical repeaters and optical switching nodes in our networks in order to obviate the need for inefficient opto-electronics conversions. Technologies that can currently only be construed in the lab will be implemented in the form of optical repeaters for commercially operated submarine cables by the late 90s. Optical switching systems will be available by the end of this century.

Telecommunications and its applications are one of our vital resources and, as such, are increasingly influencing the actions of private individuals and enterprises. This trend is intensified by the growing mobility of our society. For telecommunications operators, it is now important to find the right solutions to meet these challenges. We will be required to make telecommunications more communicative. Telecommunications cannot, and should not replace personal contacts. What it can and should do, however, is to make telecommunication more personal. The technical possibilities in the coming years will provide an opportunity to improve the quality of voice telephony, accelerate the exchange of data and make moving images possible. We will be measured by our ability to translate technical possibilities into practical applications for our customers.

4

The Japanese scene

I. Toda, Executive Vice President, Nippon Telegraph and Telephone Corporation*

This paper discusses Japanese telecommunications in the 21st century from the viewpoints of regulation, applications, services, and networks. The conclusion is that the role of networks will change from serving as infrastructure for the telecommunications to serving as infrastructure for the information society. Ubiquitous availability of broadband ISDN, intelligent networks, and possibly, all-optical networks will be needed to attain this goal. The key strategies that NTT is relying on to introduce such networks are also described.

4.1 INTRODUCTION

In the 21st century, improving global symbiosis and quality of life will be an important social paradigm. Telecommunication systems will provide one of the most important parts of the infrastructure (Toda, 1991) by greatly improving the efficiency of other systems related to the lifestyle, production, and distribution.

New applications of telecommunication systems depend on the great convenience of telecommunications as a substitute for the flows or movement of people, things, money, and information. Strategies for constructing networks that make such applications possible, and the structure of the networks and the services they will provide, are of great concern.

There are several circumstances peculiar to Japan, including unique regulations and markets and the high concentration of population and information in Tokyo. This paper offers a projection of what telecommunications will be like in the next century, focusing on Japanese circumstances.

* Currently with Fujitsu Limited.

Communications After AD2000. Edited by D.E.N. Davies, C. Hilsum and A.W. Rudge. Published in 1993 by Chapman & Hall, London, for The Royal Society. ISBN 0 412 49550 3

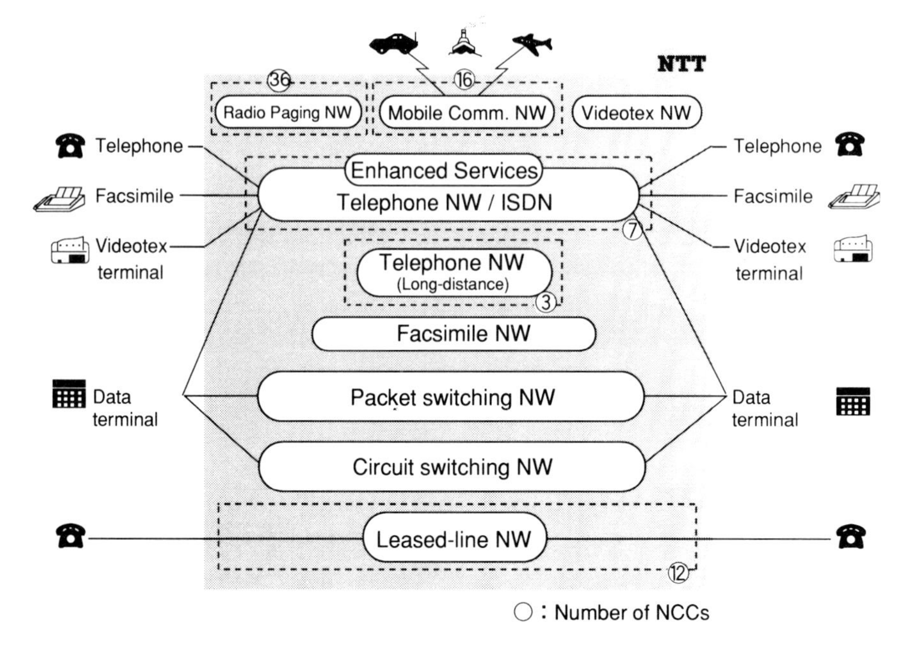

Figure 4.1 NTT business areas.

4.2 JAPANESE TELECOMMUNICATION MARKET

4.2.1 Number of new common carriers

1985 was a landmark year for Japan's telecommunications market; the market was deregulated and NTT was converted from a government-owned monopoly to a private company. This was the first step towards a sound telecom market for fair and competitive service offerings for the 21st century.

Once the market was opened, a host of new entrants rushed in. These are called the new common carriers (NCCs). At the end of 1991, NTT was sharing the market with three NCCs that offer domestic long-distance service, seven that provide regional telecom services, and 16 that provide mobile communication services (Figure 4.1). KDD, the established international carrier, shares with two international NCCs (MPT, 1992).

Under the provisions of the new Telecommunications Business Law, common carriers that possess their own transmission facilities must obtain permission from the Ministry of Posts and Telecommunications to start or end business operations, and they must file tariffs for authorization. In addition, NTT has to submit its annual business plan, distribution of net profit, and changes in officer appointments for approval. Aside from these requirements, the common carriers are basically free to pursue any business activities they want.

Since the market was opened to free competition, NTT has cut its long-distance service rates every year by an amount equivalent to more than ¥100 billion reduction in annual revenues. At present, NTT's long-distance service is still profitable, while its local call operations are perennially in the red.

4.2.2 Market share of the NCCs

Concern for the viability of the NCCs has led the Ministry of Posts and Telecommunications to grant the long-distance NCCs access to NTT's local networks without having to pay contribution charges. This means interconnection call charges are equivalent to NTT's ordinary dial call rates. In the past the Ministry approved long-distance tariff rates for NCCs that undercut NTT's rates by about 20%. As a result, the combined NCC share of the domestic long-distance market in fiscal year 1991 reached 16% in total and close to 50% along the Tokyo-Nagoya-Osaka corridor, the prime market region, just three years after the NCCs started operation. Similarly, international NCCs have captured about 10% of the market.

Since the objective of creating a competitive market has nearly been achieved, the NCCs have agreed to revise the existing interconnection arrangements in 1994. We assume this should lay the groundwork for fair and competitive telecom market conditions well into the next century, and that competition will now shift from rates to quality and range of services.

4.3 APPLICATIONS

How telecom networks evolve up to the turn of the century and beyond will largely depend on how people use telecommunications and how telecom applications affect our lifestyles and ways of doing business.

4.3.1 Telecom applications

We can consider the key ways that telecom networks will be used in the near future and beyond from various angles. One important angle is the great value of telecommunications as a convenient, inexpensive substitute or alternative for physical movement. This will also help in solving the problems of an ageing population and global environmental issues. More specifically, how future telecommunication networks will be

Table 4.1 Flow of people

Daily life	Conversation
Entertainment	Movies, concerts
Office	Conferences, commuting
Service	Sales, education, medical care
Culture	Museums, libraries, religion
Government	Voting, administration, environment monitoring

used to substitute for the physical flow or movement of people, things, money, and information should be stressed.

Substitution for the flow of people

Table 4.1 lists some activities that usually require people to move from one place to another. Telecommunications has the potential to allow us to engage in these activities without changing location. For example, by the 21st century, we can anticipate a much expanded range of services and applications to support entertainment, business activities, personalized remote education, visual medical information exchange, cultural activities, government administration, and a host of other applications.

Substitution for the flow of things

Similar efficiencies can eliminate some of the need to send or deliver physical objects the way we do today (Table 4.2). We have not yet seen many examples of this kind of substitution, but they are certain to appear.

We can expect that many products that normally come to us by physical distribution today – audio-video entertainment, mass media, mail, and so on – will be available over telecom networks in the years ahead. A number of thorny issues will have to resolved first, of course, such as the impact this would have on the media packaging industry that produces tapes, compact discs (CDs), and so on. CD and video shops are pervasive

Table 4.2 Flow of things

Mass media	Newspapers, magazines, books
Entertainment	Videodiscs, CDs, game software, electronic still camera
Postal service	Mail
Transportation	Car navigation, Door-to-door delivery
Distribution	POS

Table 4.3 Flow of money and information

Broadcasting
Database retrieval
Advertisement
Business operation and management

in Japan, and their vital interests would certainly be affected if their stock in trade could be readily distributed over telecom lines.

Telecommunications can also offer a very efficient means of logistic control, such as using electronic data exchange systems to support intermodal and co-operative transport. Japan's door-to-door delivery system has traditionally been very responsive, but it will become more difficult to maintain even prevailing standards without telecommunications as the labour supply continues to shrink and roadways become more and more congested.

Substitution for the physical flow of money and information (Table 4.3)

Most fund transfers are already carried over telecom networks. As for information, companies are keenly aware of the great competitive value of information resources. More and more private corporate networks will be constructed to move this kind of information. One interesting current trend is that centralized corporate networks are giving way to distributed and more personal network configurations. One last, if controversial, issue that should be mentioned in this context is whether Japanese telecom networks will be allowed to carry video broadcast traffic in the future.

4.3.2 Trends in telecom applications in the years ahead

Most importantly, the range of telecom applications will expand to far beyond what is available today. This will make it more important than ever that telecom facilities be available and ubiquitous to serve as a communications platform for all types of social activities. To a large extent, this application-ready platform will be a prerequisite for social and cultural advancement.

Also, we should not be surprised to see communications networks evolve into a much broader information infrastructure at the onset of the next century. This transformation will be driven by the probable fusion of communications and video broadcasting and the likelihood that some products currently handled by the media packaging industry will become available over telecom lines.

Of course, current regulatory constraints would have to be loosened before telecom operators are allowed to distribute broadcast video over telecom lines. Also, prevailing notions of intellectual property rights would have to be modified before programming and other kinds of packaged media could be provided over networks.

4.4 SERVICES

To support the new types of applications, a wide range of advanced telecom services will be required of the 21st century networks. These services will have to be far more sophisticated than those available today.

4.4.1 NTT's VI&P vision

To address customer needs and expectations, NTT has a long-term service vision, whose goal is to provide services that are more visual, more intelligent, and more personal by the dawn of the new century. We call this our VI&P vision (Tachikawa and Sano, 1990).

4.4.2 Creation of enhanced telecom services

Adoption of the next-generation services on a wide scale requires significant enhancement of network capabilities; the features and functions of customer premises equipment will also have to be greatly improved. Tomorrow's monitors, for example, should be capable of displaying graphics, photographs, and graphic images, and should support a wide range of multimedia functions. This abundance of information will be displayed in a rich palette of vivid colors at very high speed.

Looking further out on the horizon, image recognition technology will open up all sorts of intriguing possibilities, such as remote recognition of 3D objects and remote counting of people.

In the realm of services *per se*, the main selling point in the growing competition among carriers to win customers will be the quality of services, that is, reliability, security, range of service options, fast restoration of service after a failure and instantaneous provision of a service after it has been requested.

4.5 NETWORKS

The following describes what we think the network will look like at the beginning of the next century and surveys the strategies we are relying on to realize that vision.

4.5.1 Network structure

Basic structure of the network

Japan is a long, bow-like archipelago that stretches over some 2000 kilometers. Essentially the country's telecom network consists of a backbone trunk that runs down the full length of the country with feeders branching off either side like a fish bone. By the 21st century, the backbone trunk will be a fibre-optic highway that can move traffic at rates in excess of a terabit per second. The number of telephone subscribers connected to the network should increase to about 70 million by 2005 from 54 million in 1991.

Tokyo metropolitan area

Tokyo is situated about halfway down the arc of Japan and is home to roughly 30 million people, or 25% of the country's population. The population density in the central wards of Tokyo is 13 800 people per square kilometre. This is much denser than the 4300 for London, but not nearly as dense as Paris, which has a concentration of 20 700 people per square kilometre. Tokyo is also very much the centre of the country in terms of telephone traffic, with some 60% of calls originating or terminating in the city. If present traffic trends continue, Tokyo is going to need major new telecom facilities even before the end of this century.

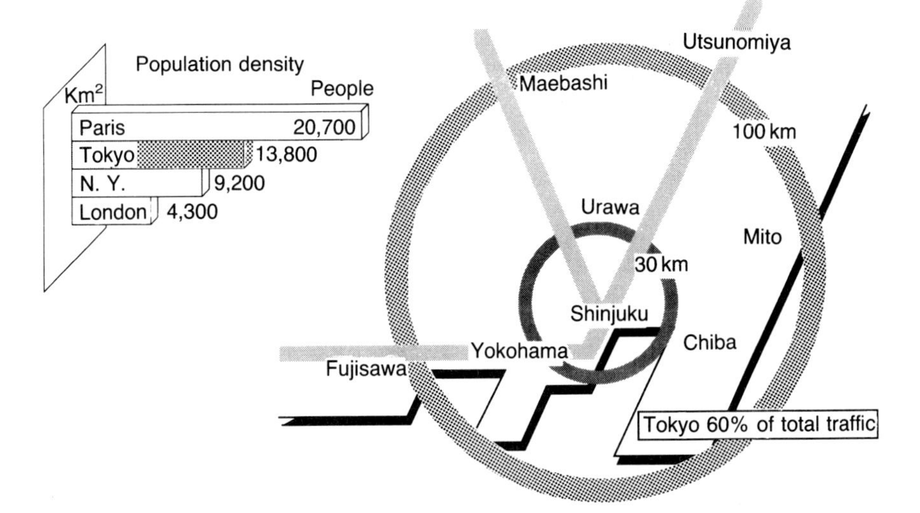

Figure 4.2 Greater metropolitan network.

Tokyo is densely built up, so it is virtually impossible to do extensive tunnelling or excavation. Thus, we are now constructing a special very large-scale network called the Greater Metropolitan Network in the suburbs. The new network consists of two giant loops – one 30 kilometres out from the city centre and the other 100 kilometres out – with feeders radiating out from the city (Figure 4.2). This arrangement permits domestic through-traffic not destined for Tokyo to be easily routed around the city and it improves reliability in the event of earthquakes.

4.5.2 ISDN

Number of N-ISDN subscribers

NTT began providing narrowband ISDN (N-ISDN) services in 1988. By the end of 1991, 80 thousand lines were in service, counting Primary-Rate-Interface lines as ten Basic-Rate-Interface lines. By 2005, there should be approximately 20 million lines in service.

N-ISDN deployment strategies (Figure 4.3)

The NTT deployment strategy for N-ISDN centers around two basic ideas. One is to keep rates as low as possible. NTT offers a single ISDN 64 Kb s^{-1} B-channel circuit at essentially the same rate as a conventional telephone circuit. Also, as ISDN lines can carry all types of communi-

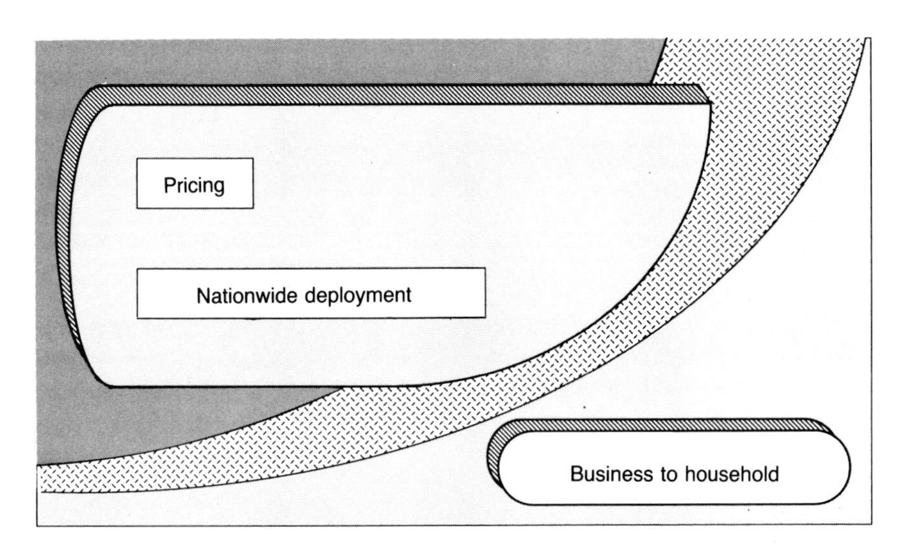

Figure 4.3 N-ISDN deployment strategies.

cations – voice, data, and image – we have implemented a common, service-independent rate structure based on actual number of transmitted bits. However, the ISDN packet service is billed by number of packets.

The other aspect of our strategy is to first target corporate customers, who have prodigious data movement needs, and then later pursue the household market. Our goal is to support enterprise-wide backbone networks as soon as possible, whether they are nationwide or local. NTT is in a favored position to follow through with this plan because our service area embraces the whole country, and also because Japan is a compact country with a land area about one and a half times that of Great Britain.

One interesting demand for N-ISDN in Japan is for supplementing the circuit capacity of existing cables. Along with the great proliferation of electronic communication devices in offices and facsimiles in homes, the demand for more circuit capacity in existing office buildings and houses is increasing. ISDN allows us to meet this demand without additional cable, so this has proven to be an effective way to accelerate deployment of ISDN.

B-ISDN deployment plan

Broadband ISDN (B-ISDN) services are scheduled to begin in around 1996, and we expect to see about 5 million lines in service by 2005. A decade later, the number of lines should be close to 40 million.

B-ISDN deployment strategies (Figure 4.4)

As with narrowband ISDN, the strategy is to target the business sector first, and then broaden the customer base to smaller businesses and residential customers. Initially, companies will start to use B-ISDN lines for high-speed digital links interconnecting their local area networks. The first wave of customers will be drawn by B-ISDN's high-speed digital transmission followed by its image transport capabilities.

If these assumptions are correct, our scenario of B-ISDN deployment will develop in the following way. First, the network must have asynchronous transmission mode (ATM) trunk lines. ATM is a key technology for supporting high-speed digital leased-line facilities. Optical fibre must be deployed in local loops for analogue telephone and N-ISDN services. In parallel with these infrastructure developments, research and development has to intensively focus on B-ISDN applications.

As demand for high-speed digital and high-bandwidth image transport picks up, ATM nodes will be deployed and ATM transmission over existing optical fibre subscriber loops will be introduced, thus extending B-ISDN availability to the whole country.

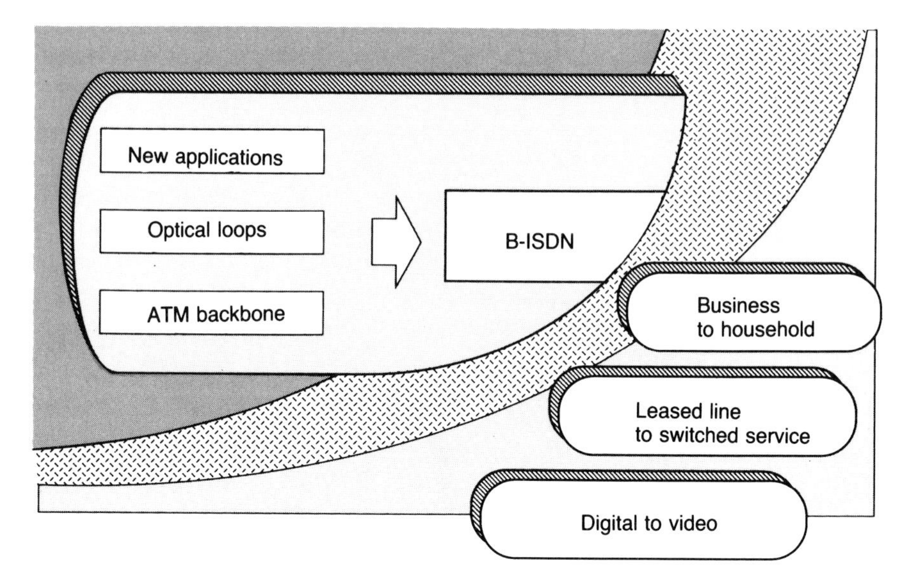

Figure 4.4 B-ISDN deployment strategies.

At the startup of B-ISDN, pricing strategy will be crucial. For example, if full-motion video and image services are to get off to a good start – especially to spark popular demand for video telephone service – the rates will have to be held down; for example, rates cannot be higher than several times the current telephone rates. Also, to put carrier operations on a sound footing, a new rate structure for B-ISDN is needed.

Transmission of high-definition and 3D images requires a bit rate in excess of $150\,\mathrm{Mb\,s^{-1}}$. To support these speeds, B-ISDN is likely to be succeeded sometime early in the 21st century by an all-optical network. By then, telecommunication networks will probably have evolved into an information infrastructure.

Optical fibre for local loops (Figure 4.5)

Installing optical fibre in local loops is a prerequisite for B-ISDN deployment. We are contemplating a phased introduction, first extending fibre to office buildings, then to the curb (a distribution pedestal serving a cluster of houses), then finally all the way to the home. We project that fibre will reach cost parity with copper sometime around 1995, assuming basic-rate ISDN distribution.

This plan is already in motion, and we have started to run fibre into office buildings and condominiums where multiple lines are installed in the same building. Fibre is also being brought closer to household users

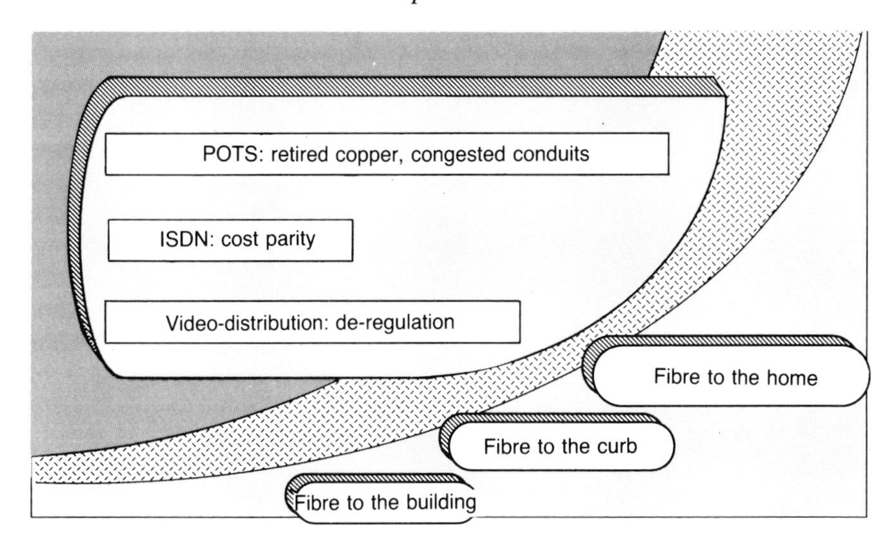

Figure 4.5 Optical subscriber loop strategies.

by a number of factors. One is that the huge amount of copper cable installed in the 1970s is now being retired and replaced by fibre. Also, many conduits and tunnels have no room for more metallic cable, and fibre, with its practically unlimited bandwidth can greatly relieve this situation. Another factor is that fibre loops will be installed to meet new demand for N-ISDN services after cost parity with copper is reached.

At present, telecom carriers are prohibited from distributing broadcast video over their lines. Obviously, if this regulatory situation were to change, this would help push down the cost of communication cable, and fibre to the home would be realized much sooner than generally expected. These various factors lead us to believe that most customers will be served by fibre by 2015.

Metropolitan subscriber networks

Major urban centres like Tokyo and the newly developed city areas, have special, intensive telecom needs. The density of computers and other electronic communication devices and the increasing use of distributed processing systems fuels a great demand for LAN and MAN interconnections, not to mention demand for connection of workstations and all kinds of multimedia terminals. Inner-city subscriber networks will thus assume a totally different kind of structure from networks in non-urban areas, especially to meet their throughput, reliability, and security requirements.

Tokyo's cellular system is a case in point. The present analogue cellular network is already very heavily subscribed, but this situation will be alleviated towards the end of this year when a second-generation digital cellular system that allocates one transmission path for an area 80 m by 80 m on the average is introduced.

Outside the city, where houses and buildings are more dispersed, mobile users will rely on satellite links in the future.

4.5.3 Intelligent network

Advanced services

NTT already offers a number of advanced services such as Free Dial and Dial Q2 (equivalent to 800 and premium call services in the United Kingdom). At present, however, each of these services is supported by its own service control point. Moreover, while service management is under the control of the end customer, the services actually have to be created by NTT personnel.

Both of these situations will be remedied in the latter half of the 1990s; we plan to put service creation capabilities in the hands of our customers, and develop a generic, service-independent service control point. Also intelligent network (IN) capabilities will not be confined to the telephone service, but will be applicable to a wide range of ISDN services.

For example, one application that shows the potential benefit to customers of the intelligent network is the simplicity of configuring virtual private networks for corporations, communities, or other private groups. To allow customers to define the groups themselves, a full set of capabilities for managing the network – in other words, customer networking capabilities – is required.

Personal communications services

Demand for personal communications services is increasing very rapidly, and we anticipate that about 20 million people will have signed up for these services by the year 2005. By then, personal communication devices will have shrunk to a Lilliputian 100 cc and personal telecommunication numbers will be assigned to people rather than to pieces of hardware. A tracking mechanism will be available even for the fixed networks, so people can receive calls at different locations if they choose to. It can be expected that notebook computers and other non-phone terminals connected to personal communication radio networks will be commonplace. Also, marrying intelligent capabilities with cellular networks will make a personalized phone secretarial service generally available.

4.5.4 Common channel signalling networks

As advanced telephone services such as our Free Dial and Dial Q2 offerings grow in popularity, and as personal communications services become available in the future, we project that the volume of common channel signalling traffic will expand six-fold, from 100 000 transactions per second today to some 600 000 transactions by the year 2000. To accommodate this great increase, we will install separate signalling networks for service control information and network operations information that is currently carried over the same common channel signalling network, and establish high-speed signalling networks with an ATM-based backbone.

4.5.5 Interoperability, integration and globalization

As the number of services, the number of service providers, and the demand for borderless interoperability all continue to increase, the importance of interoperability among public networks and between public and private networks around the world will also increase. In Japan, considerable progress toward standardized interfaces has been achieved; standard protocols for basic services covering digital cellular systems, personal communications services, and other key areas are particularly well poised for acceptance. A major standards-related issue now looming on the horizon is interconnectivity among the great number of new advanced services that will soon start to appear.

Another interesting development is that customers want to assemble their corporate networks from the best telecom services, computers, workstations and others that are available from different vendors, but they do not want the headache of making these disparate parts work smoothly together. Many customers prefer a one-stop-shopping solution, so they are turning this task over to network systems integrators, a new type of business that specializes in co-ordinating and managing multi-carrier and multivendor networks.

4.6 CONCLUSION

If the 19th century was the age of coal and steam and the 20th century the age of fossil fuels and the automobile, it is already clear that the 21st century will be the age of information and networks. Even now, the range of telecommunications applications is expanding faster than ever before. By the onset of the 21st century, we will see a fusion of communications, broadcasting, and packaged media that will change the structure of the information industry. It is expected that communications networks will cease to be considered infrastructure for telecommunications and come to be considered infrastructure for the information society.

This is definitely coming, but not until certain network developments take place – the ubiquitous availability of B-ISDN, the intelligent network, and possibly, all-optical networks. Focusing on Japan, this paper offered a projection of what the network will look like at the beginning of the next century, including prediction of some dramatic changes expected to appear in the area of network management and access networks. The key strategies that NTT is relying on to get us there were also presented.

Many corporations have already expanded their operations to span the globe. Globalization may, in fact, eventually penetrate down even to the household level. This, in essence, is what is being sought – better and more convenient interconnectivity and interoperability over one or many communications networks spanning the whole earth. For example, a roaming capability to augment personal communication services should be supported globally. For this to happen, we must promote international standardization efforts for communications services and networks to achieve greater convergence or commonality among them.

REFERENCES

MPT (1992) Announcement of new common carriers, *Information Communications Journal* (in Japanese), **10**, (2), 60–61, February 1992, (edited by the Ministry of Posts and Telecommunications), Association for Posts and Telecommunications.

Tachikawa, K. and Sano, K. (1990) The corporate philosophy and service vision of NTT, *NTT Review*, **2**, (3), May.

Toda, I. (1991) *Information technologies for the network society*, 6th World Telecommunication Forum, Geneva, October.

5

From telephony to information networking – renovating the telecommunications infrastructure

G.J. Handler, Vice President, Network Technology, Bellcore

5.1 THE AGE OF INFORMATION

Imagine the complexity involved in sending a letter from Red Bank, New Jersey, to a friend in Europe or to a business partner in Los Angeles, California without today's ability to send mail globally. Imagine the confusion and chaos that would exist without the standards of addresses, zip codes, and stamps; the flexibility of the different classes of delivery; the reliability of overnight/express mail.

It is a disturbing thought – the world without the postal infrastructure. If we agree that to qualify as infrastructure, a resource must be widely available, easy to use, and comparatively inexpensive, it becomes fairly obvious that the mail service, even with its problems, is an established infrastructure that has continually evolved to meet people's needs better. At present, I believe it would be safe to say that many people in the business environment would be quite perturbed and less productive if we did away with overnight mail. It is a service that has become quite indispensable.

The postal infrastructure has grown to accommodate people's needs on a global level. One gets quite a different feeling, however, when looking ahead to an infrastructure that has yet to be built, at current technologies that need to be integrated into the networks, at links that will be required to globally connect the networks, and at information services that will be available once this infrastructure is completed.

Such is the view of the information network of the future – the renovation of today's telecommunications infrastructure.

Communications After AD2000. Edited by D.E.N. Davies, C. Hilsum and A.W. Rudge. Published in 1993 by Chapman & Hall, London, for The Royal Society. ISBN 0 412 49550 3

We are surrounded by information – papers and books, phone calls, voice mail, and other sound-based forms, not to mention electronic text and data files, electronic mail, or information contained in still and moving images. If people and businesses are to take advantage of information, they must have the means to manage it and to have it where they need it, when they need it, and in a form they can use.

Providing these capabilities presents many opportunities and challenges for those whose business is telecommunications (transmission of information), opportunities that go far beyond the vision of telephony (transmission of sound) on which that business was built.

5.2 THE CHANGING ENVIRONMENT

For much of the past century, the goal that consumed the industry was the vision of universal telephone service, a network that would provide ubiquitous connectivity for telephone calls and access to basic voice services at the lowest possible cost. This infrastructure is optimized for 4 kHz voice in both access and switching. Customer access is through a dedicated switch connection, and services, those that do exist, are provided through the software integrated into these switches. Finally, costs are driven down through the use of a complex set of automated operations support systems that administer and maintain the structure. Over the past several decades, however, customer needs have been changing. Both social and corporate environments are becoming largely dependent on the efficient ability to create, access, store, process, and manipulate information. Information itself is being redefined to include any digital signal – voice, data, video, or image. This information is being demanded at speeds hundreds of times more rapid than in the past, and accessibility is being demanded from just about anywhere a person can go.

By taking a quick snapshot of the tremendous change in lifestyles in the US, one can easily realize the many needs for information and the many opportunities for additional technology.

The look of the 1990s is one of convenience. A person no longer needs to travel to his or her bank for cash or to pay for groceries with cash; a debit card will do the job. Debit cards and credit cards can be used for just about everything; each of these is a simple service that fits in your wallet. By 1995, there may be as many as 2.7 million point-of-sales terminals in the US, handling up to 17 billion transactions a year (Caruso, 1990).

Services that make communicating more convenient and efficient are also in demand. Services such as caller identification, call waiting, and voice messaging services are becoming widely used in the home and are also proving to reduce costs, improve customer service, and increase employee productivity in the business environment.

Telecommuting is becoming a necessity in the 1990s. People are using telecommunications to work from home and connect their offices via modem to access corporate databases, to send/receive documents via facsimile, to send/receive electronic mail messages, and to conduct business via teleconferencing. The luxury of convenience is not the only demand. California, for example, is demanding that industry establish an environment for 'work-at-home' to help preserve their ecological environment. This change in work-style may require much less travel and face-to-face interaction, but it will demand much more in digital connectivity, wider bandwidth, integration of voice and data, more advanced voice features, and increased high/low-speed data communications capabilities.

When people are required to travel, however, they are demanding the ability to take their voice and data services with them. Cellular and wireless networks currently make it possible to reach people while they are driving or walking. Cellular service in the US has grown from only a handful of installations in 1985 to an estimated five million in 1990. According to the Eastern Management Group Firm, by the year 2000, half the nation will regularly use at least one advanced form of wireless communication, whether it be a cordless home phone, cellular phone with call waiting and call forwarding or a car equipped to tell its driver directions (Star Ledger, 1992). At the start of the 21st century, we can expect to see a larger percentage of cars equipped with special data access capabilities that will enable on-line, real-time information, such as customer information, delivery instructions, directions, maps, and addresses to be viewed on special devices. An even more advanced wireless technology will permit users to communicate over broader distances as well as give them the option of having just one telephone number for their house, car, office, and cellular phones. Taxis, limousines, fleet drivers, and police vehicles will be able to communicate with each other and no longer just their dispatchers.

Video services mark incredible growth in market demand. Although we in the telecommunications industry believe that we have achieved universal telephone service, more homes in the US have television service than have telephone service today. People are just beginning to realize the potential uses for video – renting a movie and watching it at home, recording your wedding or anniversary party on video tape, recording the game while out running errands, etc. (Ninety million VCR units were sold in the US from 1984 to 1991 (Chynoweth, 1991).) Although video applications are predominantly in the entertainment arena at this time, support for applications for education, training, video conferencing, and video telephones is projected to become a multi-billion dollar market in the last half of this decade. For example, CDI™ (Compact Disc Interactive), a newly-introduced CD-based system, allows users to interact with video-based entertainment and education applications on a standard TV set.

The computer revolution is yet another example of market demand. The explosive penetration of the personal computers owes its success to people wanting to store, manipulate, and access information. (Currently, about 30% of the US households have a personal computer (Chynoweth, 1991).)

Both small and large businesses as well as the health and academic arenas are beginning to view information as a strategic asset in the 1990s. Small businesses need to efficiently manage inventory and billing information; large businesses need to connect their networks to share data and designs; hospitals could strongly benefit from the ability to share X-rays and CAT scans; and schools will be able to share expertise. The security and reliability of this information can not be compromised.

As can be seen from this brief survey of changes, opportunities, and needs in both the social and business lifestyles, society is both starving for information and is flooded by it. In this environment, an information network, which extends the reach of the customer through time and space, can be the determinant of competitiveness as well as the general quality of life. It is essential that we carefully align the characteristics of the target network to the demands of our customers in order to ensure that the infrastructure being both renovated and built will be even more promising that the individual technologies it is comprised of.

5.3 NETWORK OF THE FUTURE

The Bell Atlantic brochure, *Delivering the prose: a vision of tomorrow's communications consumer* includes the following comment.

> We envision a single device through which a consumer may watch television, view his cousin 1000 miles away, conduct his office job on a telecommuting basis, engage in home banking and shopping, compose poetry, receive messages wherever he goes, turn up the heat or turn off the stove wherever he is.

Daily, human activity is saturated with information; the different shapes and forms of information are unlimited. We, the designers and users, must take care to identify the services that the marketplace will demand, and we must co-ordinate this effort with the necessary planning for the more long-term global network evolution steps.

The characteristics of the telecommunications network of the future will be driven by business and residential market trends. The sophisticated business end-user views information and the ability to access it as critical to ensure continued corporate profitability. These end-users have made significant investments in terms of resources and have deployed, in some cases, customized information technologies that satisfy their needs. They are now more aggressively seeking information networking solutions that allow them to access and integrate information that is external to their immediate environment. The critical need is to provide information networking solutions that will tie together the business and residential environments. These 'work-at-home' capabilities will help to satisfy the change in work-style, the need to preserve ecological environment, and the desire for convenience.

A predominant characteristic of the residential end-user is the need for information access to enhance personal lifestyles and leisure activities. Products and services (whether entertainment activities, flexibility in telecommunications offerings, or consumer information guides) are becoming increasingly focused on customer convenience.

The end result is the need for an increase in the use of automation, computerization and efficient, convenient access to information. This information must be able to be transported from any format to any format (e.g., text to voice), over any distance, in any volume, and at any time.

To understand better how we can achieve these characteristics of the network of the future, it is important to have a common understanding of their definitions and of the necessary changes to the telecommunications network.

The following five characteristics summarize the predominant market need pressures for the new information networking infrastructure.

1. Universal access and transport
   The evolving environment is one having not only widely distributed information sources, but also a massive, yet dispersed, population of information seekers. Fundamental to this environment is the need to access, retrieve, and manipulate public and private information, in any form, as well as giving the information provider access to the mass market. The ability to interconnect with any information source or information transport resource must be available, and customers should not be limited by the location of the technology that provides the services or access to those services.
2. 'Just-in-time' service
   The customer demand is for information at any time. This implies responsive, sometimes instant, provisioning of carrier services, as well as transport capacity. The automated, yet still time-consuming,

methods of today's carrier-controlled operations will have to change to automation under customer control.

3. Customized services
   The wide range of information needs, coupled with the varied capabilities of customers' equipment, requires customized information networking services that may differ in characteristics (bandwidth, presentation schemes, transport parameters) and interfaces to Customer Premises Equipment (CPE) or other networks. The requirement for rapid response and customization in the network is also a result of the rapid technology evolution at the customer premises.
4. Third-party service provisioning
   With so many information sources and so many application needs, services are likely to be defined and offered by many entities. To accommodate this, the information network provider must provide capabilities which facilitate the creation and provision of services by others, including customers themselves.
5. Service security and survivability
   The concept of 'just-in-time information' requires the access to information to be secure and survivable because customers are increasingly relying on information as a critical asset. Capabilities for protecting stored or in-transit information must include protection from information thieves, as well as instant recovery from network, and even customer equipment, failures.

An infrastructure that supports the above characteristics sounds exciting and may prompt reactions of 'Can I have it tomorrow?' But these information network characteristics are a target view of the telecommunications network, and will guide the evolution of the network toward meeting the needs of information-dependent customers. If these characteristics are thought of as goals for the network of the future, the attainment of these goals will result in a network that imposes almost no restrictions on its user in terms of service availability, creation or geographic location. It is a network that provides range of choice, convenience and, from the end-user perspective, ubiquity.

5.4 TURNING OPPORTUNITIES INTO SUCCESS

Forecasting wondrous visions of the future based on a snapshot of technological opportunities, social and market trends, and people's changes in lifestyles is a stirring and stimulating exercise. However, a more practical vision comes from looking closer at the technological developments, the applications they promise to support, and the network characteristics they will deliver.

5.4.1 Universal access and transport

Over the past decade, the telephone network has been evolving into a digital network, offering the ability to integrate voice and data services. Bellcore has been working closely with end-user applications developers, equipment providers, and end-users to ensure the mass delivery of reasonably-priced and user-friendly integrated services digital network (ISDN) customer terminal equipment and applications.

The industry is extremely excited and confident that ISDN will be a vehicle for bringing new products to the mass market. A standard ISDN applications programming interface (API), MS-DOS access for PC terminal adaptors, and a UNIX™ API are all currently being designed to assist application developers, who will make the potential of ISDN a reality. US exchange carriers have committed to begin deploying ISDN this year. We expect that by 1994 most of the US regional telephone companies will have 50% or more of their lines ISDN-ready, with some regional telephone companies as high as 80–90%. With National ISDN-1 in place, residence and small businesses will be able to take advantage of low-speed data services such as those supporting 'work-at-home', desktop conferencing and advanced call management. Large businesses will profit from PBX networking and connectivity to branch locations. Internationally, ISDN is also being implemented. France Telecom (FT) launched the world's first public ISDN, Numeris, in late 1987. Japan shortly followed with its introduction of ISDN in 1988. ISDN was installed in Barcelona for the 1992 Olympics. Most other European countries plan to become significant users of ISDN by 1995 (*Data Communications*, 1991).

Millions of miles of fibre have been installed, but because of economic and regulatory constraints, fibre optic cable has yet to penetrate the local access loop in the US. Fibre-in-the-loop (FITL), targeted for 1993 deployment by many US carriers, is a technology that will increase the availability of higher bandwidth and make it possible to provide information and entertainment services on demand to businesses and residences. Although today's local loop still consists mainly of twisted-pair copper and provides only narrowband services such as POTS, pay phones, and analogue and digital private-line data services, the local exchange carriers are making rapid advances toward an optical network that would support broadband services such as remote medical imaging, high-definition TV (HDTV) and video telephones.

As the carriers are planning to deploy thousands of new fibre lines beginning in 1992, a set of national and international standards for fibre optic transmission systems is nothing less than critical. The two primary goals behind the SONET (synchronous optical network) standard are to learn how to manage the growing bandwidth and to protect it. The SONET standards are key in the evolution of an interoperable, efficient and economical fibre transport infrastructure for the public network. In addition to providing higher speeds, SONET offers bandwidth-on-demand capabilities through SONET-based multiplexing technology, which will eventually be available to end-users.

The move to SONET will require major changes in the network, both in transmission and switching (see Figures 5.1 and 5.2). The traditional hierarchical design of the telecommunications network will evolve to ring structures which provide a great amount of capacity and reliability. The first generation of SONET equipment is expected to provide capital savings and improved survivability and flexibility over current technologies.

SONET is already under trial and deployed by many US telephone companies, and most major public network equipment suppliers are developing or producing SONET equipment. SONET is also being supported on an international level. (Japan began to deploy SONET in December 1990; Germany, Great Britain, France and Italy are phasing in SONET in the 1991/1992 timeframe.) Building on this common approach, global applications will share the same capabilities and benefits of high-speed data networking.

Another obvious challenge is to replace the multiple service-specific networks that make up today's public network, and thus improve efficiency as well as performance. Such networks have separate access interfaces, switching fabrics, and transmission formats to meet the needs of voice, data, and video communication. Bellcore is working to define a broadband network that will be able to efficiently handle the diverse transport requirements for all types of information. The two key technologies involved are SONET transport (as previously discussed) and asynchronous

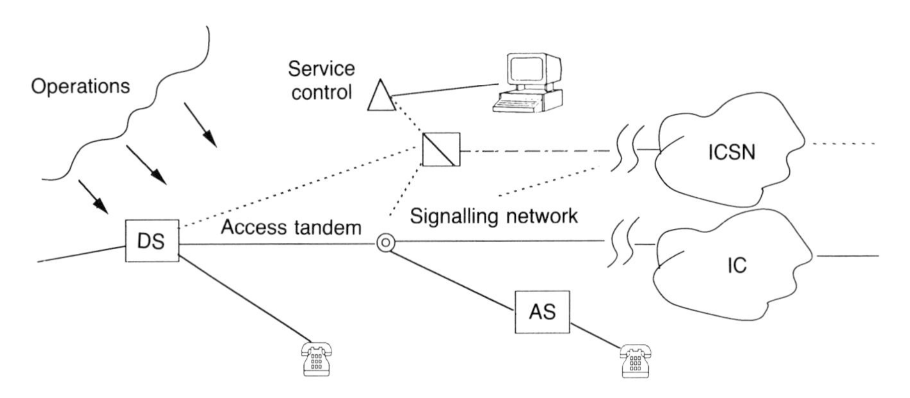

Figure 5.1 Current architecture.

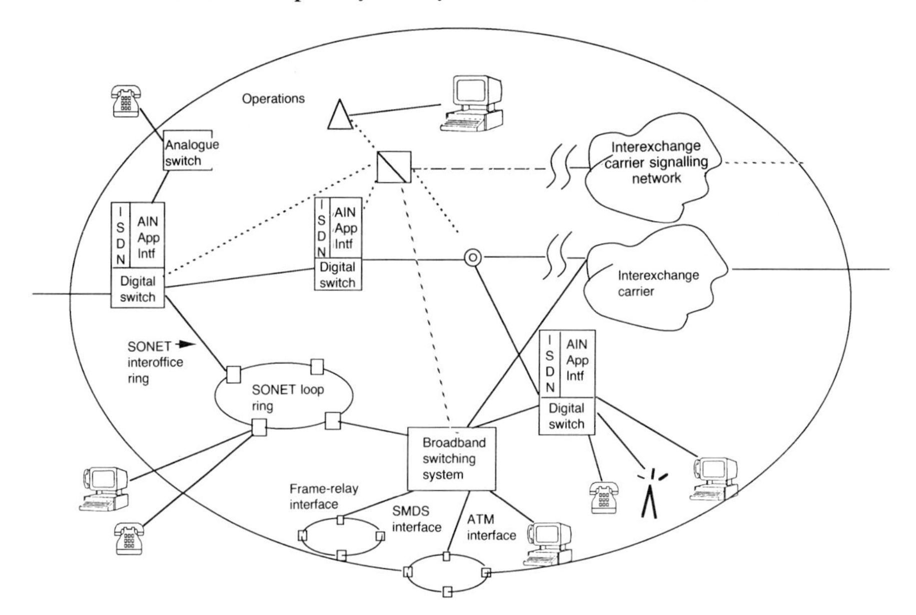

Figure 5.2 A 1995 network architecture.

transfer mode (ATM) or cell switching, a standard switching and multiplexing technique that supports both continuous bit rate (video) and variable bit rate services (file transfer), as well as both connectionless and connection-oriented services. Initial deployment of this platform (as early as 1992 and 1993; ATM products are out this year) is targeted at business customers for high-speed data transport and then evolves to support multiple types of high-speed communications (e.g., video, high-speed data, imaging).

5.4.2 'Just-in-time' service and customized services

As the network evolves to accommodate high-speed communications of all types of information, the supporting platform must evolve to intelligently provide and manage these services. In order to provide rapid service provisioning, increased automation and integration of operations systems are necessary. This will allow, for example, customers requiring telephone service at a new home or vacation home to order and receive dial-tone instantly. In addition, Bellcore and its client companies have been working on the advanced intelligent network (AIN), a service- and vendor-independent platform that will eventually allow customers to develop and deliver new voice and information services, customized to their individual needs. AIN will be introduced through a series of releases, beginning in 1992. A goal of AIN is to build upon the embedded base of the telephone companies' networks. One advantage of AIN is that suppliers will no longer be required to recode the complex switch software each time a new service is added. With the AIN platform, a new service program can be implemented quickly by the telephone companies or even by customers themselves.

But what good are these advanced features if you can't take them with you everywhere you go? Another benefit of the intelligent network is mobility of services. AIN combined with low-power digital radio will allow a person the portability of a pocket terminal, without the need for a wired connection or even a wired-in base station. The tremendous growth in sales of the smaller version of the cellular phone (briefcase or handbag size) is paving the way for portable communications. Cellular phones were designed for use in vehicles, travelling up to 60–70 miles per hour. This potential speed factor required a relatively strong signal – from 1 to 10 watts – to reach across the relatively large geographic cells the user will be passing through. This power requirement is not a problem when the phone can rely on the electrical system of a car, but the battery size would be inconvenient for extended use while walking around.

Telephones for personal communications will be used at walking speeds allowing for smaller geographic cells, and therefore will require much less power. A smaller, lighter, longer-lasting power source (about a hundredth of a watt) will be used, thus making the handset comfortable enough to be carried in the pocket of a shirt or jacket. The market for Personal Communication Services (PCS) is estimated to be as much as $10 billion within just a few years. By 1993, enough metropolitan areas could have radio ports in operation to be able to support a cordless version of the public telephone (Brush and Butler, 1991).

5.4.3 Third-party service provisioning

As we move to an information infrastructure that enables third-party service provisioning, we need to design a new way of organizing software in the network and its operations. As mentioned in the previous section, AIN will eventually make it possible for customers to develop and use new voice and information services, customized to their needs. These service creation capabilities in AIN can begin to provide the opportunities for implementing third-party services, but the current environment, based on separate network software and independent operations support software systems, makes it difficult to support sophisticated service creation well.

To push third-party service creation and provisioning further Bellcore, in collaboration with its clients and industry representatives, has been working on a concept called information networking architecture (INA). A key element of this architecture is a distributed processing environment (DPE), which allows applications to communicate easily. Such applications could reside in the network or be offered by value-added providers outside the network. The DPE would allow for easy addition of new applications, as well as encourage co-operation among many applications. Of course, national and international standards will be required to implement such an environment.

INA is being developed through a series of industry experiments. An important element in this approach is collaboration with *all* stakeholders involved, including network equipment and terminal suppliers, the computer industry, independent local exchange carriers, inter-exchange carriers, independent service providers, value-added network providers, and users.

It is important to share the knowledge from these experiments to help support efforts around the world aimed at defining an information architecture: RACE, ESPRIT, the international telecommunication information networking architecture (TINA) effort, to name just a few, as well as network and operations systems suppliers' evolution toward platforms which implement such software concepts. Unprecedented co-operation will be required for timely solutions.

5.4.4 Service security and reliability

FITL and SONET, discussed above, will provide increased service security and network reliability. In particular, SONET will improve reliability in a couple of aspects.

- By simply reducing the number of network elements, the reliability of the network increases.
- SONET's embedded operations capability improves end-to-end network management and control, increasing network reliability and survivability. SONET rings being deployed this year are self-healing.

The call control capabilities of AIN will allow for sophisticated access screening to provide added security. Clearly, all new network capabilities must include security consideration.

5.5 CHALLENGES

Customers need access to information. They have already invested in sophisticated hardware and software that can facilitate local management and processing of information. Information service providers are ready to make information available, to process it, and to help customers to manage it. By defining the characteristics of a target architecture and then identifying a series of transition steps, we can move closer to such a network and seize the many market opportunities.

There are many challenges ahead of us. Several exceptional implementation challenges are involved in moving from the current software environment to a distributed processing environment. Structured programming techniques have never been scaled up successfully to a task this size. Distributed processing has not yet been tested for survivability and fault tolerance when attempted on such a large scale. In addition, software productivity and quality gains are necessary ingredients for real, timely progress. It is also not clear how current network and operations systems can be evolved toward such a distributed structure. We anticipate the experimental approach being used for INA will uncover solutions to many of these problems.

The economic issues go beyond the traditional business domains of the telephone companies, bringing in issues and stakeholders from the computing industry, government (e.g., the Internet), cable TV and so on. There is little precedence to give us comfort that regulatory bodies are capable of dealing with this complexity in a way that stimulates (or even allows) the development of an information networking infrastructure. Yet the market will demand this infrastructure. We can help: as the designers and users of the information network, we must be willing to take the success stories of research labs and aggressively seek industry commitment to deploy them.

Another challenge is to answer the threats and accusations that this infrastructure will cause us to be smothered with mountains of irrelevant information, will ultimately dehumanize people, or will widen the gap between rich and poor. We have to be careful to seek out those opportunities and technological discoveries that will answer the demands of different markets and will provide useful services to customers. Similar to the postal infrastructure and to universal telephone service, we must coordinate our efforts and connect our networks to meet the global needs. The result will be a global information infrastructure that realizes the full potential of information, to the benefit of all.

REFERENCES

Brush, Gary G. and Butler, Christine H. (1991) A more personal kind of communication, *Bellcore Exchange*, September/October, 21.

Caruso, Rich (1990) Network services: the revenue opportunity of the 1990s, *Bellcore Exchange*, May/June, 4.

Chynoweth, A.G. (1991) *Towards multi-media services on broadband networks*, November.

Data Communications (1991) Public networks '91: a world atlas, October.

Star Ledger (1992) Firm predicts surge in use of wireless gadgets, January 5.

6

Social implications

I. Miles, PREST, University of Manchester

6.1 PROLOGUE: JANUARY 1ST 2001

Forecasting is a difficult art at the best of times, and it is likely that we are not in the best of times at present. There is good cause to believe that – in addition to the major geopolitical transformations of the early 1990s – Western societies are in a period of structural change. In such periods we cannot rely upon extrapolations from what we already know. Some long-term trends may run up against limits and countertrends; and while many trends are liable to persist in quantitative terms, there may be substantial shift in their qualitative significance.

Though the diagnosis of qualitative change and turning points in social affairs is always contentious, it can be helpful to see sociotechnical development as proceeding through various stages – what we shall term 'socio-technical formations'. This has substantial implications for how we go about forecasting. Forecasting *within* a formation can rest on assessing the consequences of the further elaboration and diffusion of practices and relationships that are already established. But in looking *across* formations, such an approach may wrongly project forward features of the present day into the future. Activities and artefacts may come together in new ways, small changes may accumulate to form a new gestalt. Grasping the prospects for a coming stage of development may require examining problems apparent in the current stage which need to be overcome; looking at vanguard groups and organizations who are seeking to operate in new ways; and/or examining what social pressures and technological opportunities are likely to present themselves.

The task of thinking ahead in circumstances of structural change is bound to be even more difficult if we are dealing with more than one

Communications After AD*2000*. Edited by D.E.N. Davies, C. Hilsum and A.W. Rudge. Published in 1993 by Chapman & Hall, London, for The Royal Society. ISBN 0 412 49550 3

such change. To talk of prospects for the 21st century is exceedingly challenging. If the last two centuries are anything to go by, sociotechnical formations are liable to succeed each other at periods of roughly 40–50 years. In the communications field itself, traditional models of regulation have broken down, and new frameworks are emerging. Technological and organizational change have stimulated new demands for communications functions and new strategies for communications management; they have led to the erosion of boundaries between media and of old assumptions about the structure of communications costs. The future role of communications will be forged within the context, furthermore, of dramatic shifts in some of our fundamental perspectives on human life such as are liable to accompany new 21st century knowledge concerning natural and artificial intelligence, biotechnology, space technology and environmental systems. It may be possible to identify extremely broad trends in human activity, but the fine texture of life in the remote future is hard to discern and is bound to be shaped by a myriad unpredictable events. Let us not seek to envision the whole of what is bound to be a tumultuous time in human history, but instead restrict ourselves to the nearer future – to January 1st 2001 and the years immediately following it.

6.2 COMMUNICATIONS AND SOCIAL CHANGE

The history of the development of modern industrial societies is also the history of communications technologies. Means for the transportation of people, goods and information were vital features of industrialization and the emergence of modern services. They permitted the introduction of extensive divisions of labour, with the high levels of co-ordination of production processes and linkage of production to markets that are crucial to modern economies. The geographical spread and spatial organization of living and working environments reflects the growing communication capacities of our societies, as their populations and economic activities have burgeoned over the last two centuries.

It is not only economic activity that has incorporated telecommunications. Bearing in mind the uneven diffusion of telephones in the UK today, it is unsettling to read the following advice in a manual on child health:

> In order for you to play your full role as the primary provider of health care for your child, little specialist equipment is required in the home. More important than any medical equipment is the establishment of your network of communication with your family doctor and health visitor, local child health services and your local hospital.

> Having a telephone at home, or access to a phone, is then your most important piece of health care equipment . . .
>
> (Baum and Graham-Jones, 1989, p. 6)

Early developments in improving communications concentrated on extending and improving transport infrastructures, enabling more rapid physical movement of people and messages in the forms of letters, newspapers and other texts. With telegraphy, texts could be communicated near-instantaneously – the main delays related to the operators encoding and decoding of messages. Telecommunication of normal speech became practicable, with telephony and radio (mainly facilitating one-to-one and one-to-many communication flows, respectively). These latter innovations diffused rapidly to businesses and households, and have been joined more recently by video broadcasting, text broadcasting (teletext etc.), new forms of text telecommunications (fax, videotex, electronic mail, etc.) and, to a more limited extent, video telecommunications (where PC-based videophone/videoconferencing systems look poised to take off). These telecommunications and broadcasting facilities have become evermore pervasive as the relevant peripherals have become cheaper and smaller, and gained new functionality by way of memories, portable and mobile devices, and so on. Continuing change in telecommunications capabilities can confidently be predicted.

There is surprisingly little systematic social research on the long-term implications of evolving communications capabilities. There has been some extremely interesting historical research, such as Beniger's (1986) *The Control Revolution*. But more future-oriented studies have in the main consisted of sweeping accounts of the emerging 'information society', or somewhat self-serving accounts of the benefits of new systems that might emerge from support for R and D. (A superior example of the genre is the set of PACE reports, generated as background to the European Commission's RACE programme for advanced communications (e.g., PACE, 1992).)

Looking at this futures-oriented literature, and scrutinizing expert viewpoints, we have concluded that much of the diversity of views about the implications of IT-related change, and the nature of the emerging 'information society', has to do with two sets of attitudes. (Cf. the literature-based studies of Miles (1988), Miles *et al.* (1988); a survey-based study which indicated a similar structure of expert viewpoints is Rush *et al.* (1986). Efforts are currently being made to incorporate this analysis into a computer-based support tool for scenario analysis; cf. Ryan & Aboulafia (1992).)

The first of these sets of attitudes concerns whether this change is (a) profoundly important – of substantial speed and scale, affecting almost all aspects of social life in a revolutionary way or (b) more of a gradual evolutionary change, proceeding more slowly and unevenly due to a large number of organizational obstacles. Are the new communications technologies underpinning a shift to a dramatically new form of society, in which our ways of living and working are thoroughly transformed – and along with them our values, class structure, and much else – or are they best seen as no more than a further step along a well-established route, which may bring new phenomena but is unlikely to mean fundamental change? In the former perspective, 'information society' marks a new stage in human civilization; in the latter, the term is rather hyperbolic, since all societies have been dependent on information, throughout human history.

A second important set of attitudes concerns whether these technological changes are viewed (a) in a basically optimistic fashion, as contributing to freedom and human potentials, or (b) in a more pessimistic way, as being used for greater social control. Are communications technologies liberating – as is suggested by many commentators who argue that their role in the downfall of Soviet and East European communism was crucial – or do they serve as new means of social control, paradoxically alienating people further from one another? The two perspectives differ as to whether new communications facilities will be used in 'information society' to increase mobility and meritocracy, or to reinforce privilege and information inequalities.

While various intermediate points of view can be discriminated, in order to highlight contrasting beliefs and policy prescriptions it can be helpful to treat these sets of attitudes as two dichotomies. The four extreme viewpoints generated by combining these dichotomies (Figure 6.1) can be used to classify much of the literature. They indicate distinctive understandings of the relationships between social and technological change. Scenario 1 corresponds to many of the optimistic, even Utopian, writers about information society. These authors argue that we are moving into a completely new civilization based upon access to information resources – 'computopia', the 'post-industrial society', the 'third wave', and so on. Scenario 2 also projects profound change, but stresses the potential of new communications systems for authoritarian surveillance and control – *1984* was delayed for a few years, and acquired *Brave New World* attributes, so as to appear more as a critique of affluent transnational capitalism than of austere state socialism.

Scenarios 3 and 4 represent the views of commentators who are more sceptical as to claims about the revolutionary nature of new IT. New technologies are seen as just another step along a trajectory that is taking us toward greater liberty and enlightenment, on the one hand, or more manipulated and deskilled societies on the other.

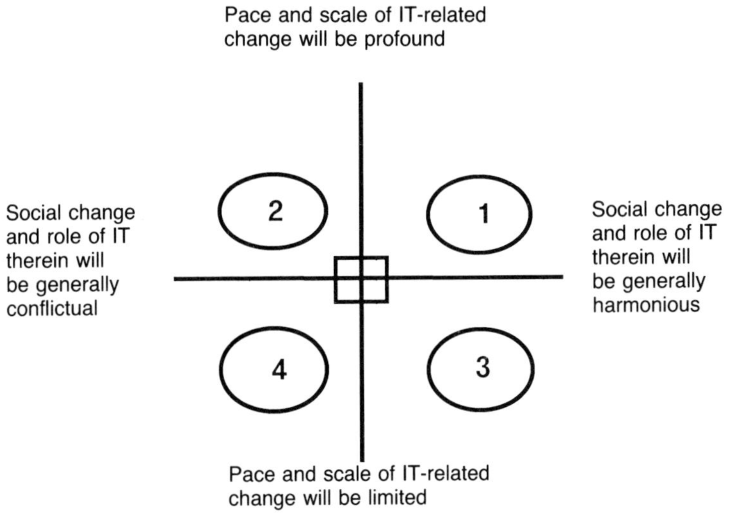

Pace and scale of IT-related change will be profound

Social change and role of IT therein will be generally conflictual

Social change and role of IT therein will be generally harmonious

2 1 4 3

Pace and scale of IT-related change will be limited

Figure 6.1 Four views of the future.

Each perspective is advanced by serious commentators (though some attract more publicity than others). Each probably has something valid to contribute – so that the debate across the viewpoints produces richer insights than any of them alone could do. More speculatively, it is plausible that the future will involve some combination of the elements of each – the open question is what proportion of each, and synthesized according to which core principles.

Moving away from the 'futures' literature, two important approaches to understanding long-term socioeconomic change, developed within recent social research, provide alternatives to seeing the future as either a continuation of the present, or as a new form of society altogether, as either Utopia or dystopia. From innovation studies have come accounts relating major technological revolutions to Kondratieff waves and to successive 'socio-technical paradigms'. From French political economy has come the 'regulation school', with its account of successive 'regimes of accumulation'. Both accounts see industrial societies as having evolved through a number of phases, marked by a combination of discontinuity and continuity; and both identify the late 20th century as a period of transition between phases. The information society of the early 21st century appears, from these perspectives, as a further step in the evolution of industrial society. (For an introduction to the two approaches see Miles and Robins (1992); also see the collection edited by Dosi *et al.* (1988).)

The first of these two perspectives is most immediately relevant to our present theme, since it focuses upon social responses to the emergence of radical new 'heartland technologies'. These are technologies which enable operations which are common to large swathes of economic and social activity to be carried out in substantially improved ways, for example by allowing substantially lower costs, faster speeds, higher convenience, etc. As the opportunities presented by these innovations are realized on an increasingly wide scale, so new products and processes are created; new skills, organizational structures and management practices are evolved; and new industries and interindustry linkages are consolidated. Innovation researchers have charted out these processes in a large number of studies, and considered new information processing and telecommunications systems are central to current developments. However, their focus tends to be on business practice and macroeconomics, with little attention to everyday life and consumption patterns.

The 'regulation school' has, in contrast, emphasized the need for integration between styles of production, labour markets, and consumption within industrial societies. For example, the emergence of mass production implies the development of mass consumer markets: different parts of societies have to be related together. This approach has, unfortunately, not been very much elaborated to deal with issues of technological change and communications technologies, though it may well have insights to offer as to the sort of social change which is liable to provide a context for ongoing technological change.

The innovation research approach identifies the characteristics of information society in terms of the transition between what we shall here label 'sociotechnical formations' within industrial society. (The term 'paradigm' is often used in the literature, but since it has been applied to a range of very different types of technological change, there is considerable potential for confusion in continuing to use it here.) Seen as a stage in the development of industrial society, 'information society' is the sociotechnical formation associated with the widespread development, diffusion, and deployment of the new heartland technologies of IT. These new technologies have brought dramatic and continuing decreases in the costs, and increases in the capabilities, of information processing and communications of many kinds. This opens up many opportunities for new applications of IT, for new ways of carrying out familiar activities, and for new activities in which to engage.

Many social scientists, themselves steering clear of analysis of relations between social and technological change, will accuse such an approach of involving technological determinism – i.e., of explaining social change as a consequence of technological change. But this accusation is misplaced. We do not have to see technologies as developing under their own momentum or as having 'impacts' upon passive human beings.

Rather, we can see advances in technological knowledge and artefacts as changing the *opportunities* that human agents perceive themselves to be confronting. These opportunities can be seized – if the resources are there to do so – to pursue one's interests. The further development of technology is the result of managers, R and D workers, and scientists pursuing particular types of opportunity. The application of technology – which may influence the further development of technology through market and other feedback signals – is the result of users pursuing other types of opportunity, reflecting their own goals and interests. Given this complicated interaction between social and technological change, we prefer to talk of the 'social implications of technology' rather than of 'social impacts'.

As Figure 6.2 indicates, we need to introduce a number of additional concepts in order to think about the nature of the social implications of technological change. First, different human actors, located within different institutional structures and with different cognitive capacities and goals and interests, may perceive technologies in terms of different sets of opportunities. The opportunities that users identify may well diverge from those of the original innovators – and these may open up new areas of application of the products (e.g., computers shifting from number-crunching to text and image processing applications), some of which may actual run counter to the hopes of innovators and suppliers (e.g., the perpetration of computer crime). Human societies are diverse,

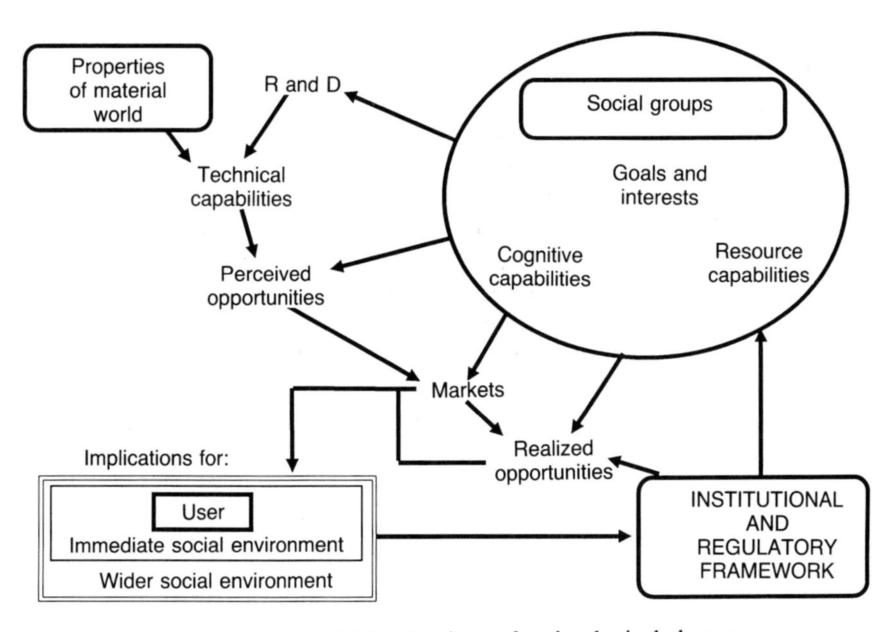

Figure 6.2 Social implications of technological change.

containing diverse, competing, and sometimes conflicting interests. The implications of technological change result from the interaction of these diverse groups, and the outcomes of such interactions will reflect the material and cognitive resources of groups pursuing different ends.

Furthermore, the use of new technologies (and the development of markets for innovations) always takes place within an institutional and regulatory context, though this may be more or less overt or covert, proactive or reactive. Social groups are defined in part by their relation to these institutional structures. And these structures may themselves change, just as everyday activities may change. We can distinguish between change that involves *assimilation* of new practices within a largely unchanged institutional regime (i.e., incremental modification of the framework), and that which involves *accommodation* of the regime to a set of radically new practices (i.e., transformation of the framework).

Second, we should always expect unanticipated consequences, even if we cannot hope to predict their content. These may arise from the intersection of a variety of efforts to apply to new technological potentials. But equally, the opportunities that are realized may well diverge from those that were initially perceived. Over time, it is likely that the opportunities associated with new technologies will come to be seen in new ways. Earlier views of the nature of the new products, and the sorts of activity which they foster, are liable to evolve. The product may well be 'reinvented', to use a term from innovation theory, and users are liable to learn more about what they can expect if they do particular things. Patterns of use thus evolve, and are usually difficult to preordain.

Third, the implications of technological change extend beyond the user. There may be other people that the user hopes to influence through use of the innovation – for example, potential consumers that could be reached by computerized telephone sales calls, potential clients about whom database information on credit-worthiness has been accumulated, etc. The user's own social position may be affected by use of the innovation (for example, an innovative firm may gain market share); the availability of traditional goods and services may be influenced by competition with new goods and services, potentially disadvantaging non-users of the innovation (as lovers of vinyl LPs plaintively note with respect of CDs). There may well be other consequences of adoption of innovations on the wider social environment – videorecorders and satellite TV mean that fewer of us will be watching the same programmes on any particular evening, so they may no longer be good shared topics of conversation, for example.

These points suggest that the characteristics of the new sociotechnical regime are difficult to predict in part because (a) they are the product of the action of human agents, (b) whose efforts to apply new technologies to their objectives may not be altogether compatible, and (c) who will take part in individual and institutional learning processes in which resolutions of competing strategies, and awareness of the 'fit' between technological possibilities and social goals, are achieved.

6.3 NEW COMMUNICATIONS OPPORTUNITIES

New opportunities are being presented by the technological trajectories of communications systems. Earlier generations of communications technology contributed decisively to earlier sociotechnical formations, making possible the characteristic forms of industrial society as we know it. The new communications potentials now becoming available are likely to be used in equally significant ways. The sociotechnical formation we call 'information society' will result from social actors perceiving and acting upon the opportunities associated by the new technologies.

Traditional communication essentially depended on the parties involved being in the same place, at the same time, using a common set of signs and words, and with restrictions on the number of participants. But we are being moved into a more fluid situation, where (at the extreme) anyone can communicate with anyone or any combination of people, anywhere and any time, using any combination of media (including the use of speech and text, graphics and gestures, even video imagery). Keywords here include mobile communications, asynchronous communications, multimedia systems, interactivity, interoperability, intelligent networks. Increasingly smaller and more powerful devices, capable of handling digitized data of many kinds, will become available at relatively low costs; this will permit the emergence of a variety of new communication and information services, including new styles of publishing in multimedia form. Among the options available will be much more 'sensory realism' in communications, although the virtual realities that can be generated may differ in many respects from the familiar world of experience; it will be possible to communicate with and to monitor people and processes to a far greater degree, and in a vastly expanded range of locations, than is currently the case.

One implication is immediately apparent. The technical constraints on communication are being relaxed, and the technical possibilities for communication increased. What follows, then, is that social constraints may well become far more important. And one of the traditional social constraints – financial resources – may become less significant for many (but decidedly not all) forms of communication as the costs of the technology plummet. In order to prevent our drowning in a sea of data, and being deafened by a clamour of would-be correspondents, individual, organizational, and societal practices will be reshaped (Figure 6.3).

Throughout history communications have been regulated by custom

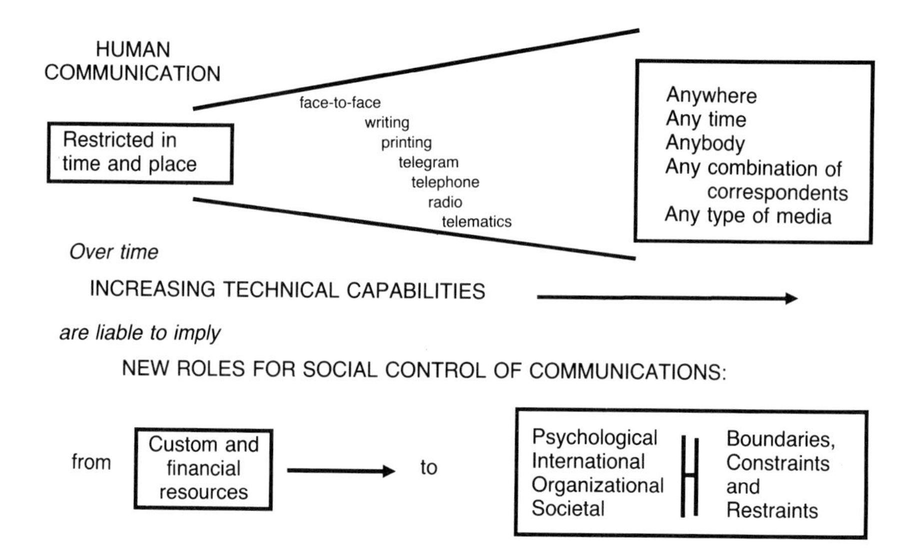

Figure 6.3 Relaxation of constraints on communication.

and convention, and access to the means of communication. States have regulated different media in different ways – for instance the UK's regulatory regimes for books and newspapers, for broadcast media, and for telecommunications. (These regimes are placed under strain as new media arrive and old media converge.) Organizations have regulated access to senior personnel by means of human gatekeepers. But now we have the potential for a communications explosion based on new media (e.g., faxcast, electronic books, multimedia), new ease of publication in traditional media (e.g., desktop publishing, home videos and camcorders), increased access to communications (e.g., cheapened and mobile communications), and new personal and organizational strategies making use of such facilities (e.g., direct mail, direct fax). New gatekeeping and boundary-setting methods may have to be introduced.

One of the great hopes of some members of the IT community is that these methods can be automated – the vision is one of 'agents' that operate for us within communications environments, using artificial intelligence to winnow out the chaff from the useful information (and perhaps to do some of our abstracting and synthesizing of data), to bar unwanted calls and schedule and prioritize other interpersonal contacts, and to help us locate the right information and human contacts in the ever-expanding web of media. This vision has considerable appeal (as well as obvious dangers), but our practices will probably evolve so as to make most satisfactory use of the new possibilities. This may mean new styles of assimilating information, and new ways of valuing information, at a psychological level. Interpersonal behaviour may change – as in the 'netiquette' that has been developed to govern exchanges on new computer media. It is plausible that new attitudes to privacy will emerge, with efforts to tighten boundaries – as in controls over junk mail and junk faxes. Organizations are likely to control access to information and personnel with a variety of human and automated ways of allocating privilege and priority in the use of services. And, as already indicated, regulatory regimes are likely to seek to assimilate change via minor reforms to their practices, but eventually to be forced to make major accommodations to the new communications landscape.

Perhaps this will be among the greatest challenges of the early 20th century – developing the social innovations which allow us to reap maximum benefits and incur minimum costs from use of the new communications possibilities. There is bound to be a large number of trade-offs among social concerns such as privacy, freedom of expression, freedom of information, and network security – and this is likely to lead to friction between social interests with differential investment in these concerns.

6.4 IMPLICATIONS OF CHANGE

The new opportunities discussed above are liable to be seized by interested social actors. Businesses and public agencies can adopt new styles of organization, by-passing existing middle management, if appropriate skills and practices can be introduced into shopfloor, front-office and field workers. New types of media can be used in education, training, and social participation, as well as in interpersonal communications and entertainment. New problems are liable to be posed, of information overload, lack of privacy, and blurring of conventional and virtual realities. We shall briefly examine the issues that are posed under a number of broad headings.

6.4.1 Work and employment

Debate has raged for over a decade on the implications of new IT for work and employment, and views have been polarized as in the quadrant of Figure 6.1. Forecasts have diverged as to whether prospective technological changes will be on balance creating jobs or destroying employment, increasing skill requirements or promoting deskilling. Views have also varied as to how significant the new technologies are as compared to other developments such as globalization of trade and production, deregulation, market volatility and new management philosophies.

A recent major study for the European Commission reviewed these debates, and analysed trends, case studies, and computer modelling. It reached numerous conclusions as to the likely developments over the coming decade, and pointed to trends to expect in the early 21st century. Results of two major programmes of work were presented in Brussels at the conference Information and Communication Technologies: Social Aspects, Employment and Training 17–18 October 1991. (A useful overview is presented by Freeman and Soete (1991), while material on quality of working life is provided by Ducatel and Miles (1991).)

Major conclusions of this study can be briefly set out. The balance between employment generation and job destruction will be a product of many forces, only a few of which are directly related to IT. (Policy stances on trade and growth, environmental issues, and the international division of labour are among these.). The EC study noted 'a strong, persistent demand for some specialised high level technical skills; a rapidly growing demand for multi-skilling; a declining demand for unskilled labour and a need for increased communication skills at all levels' (Freeman and Soete, 1991, p. 4 of the Executive Summary). Pervasive use of new IT is generally associated with increases in the average skill requirements of the workforce, though this trend is not uniform, and there is some polarization between skilled 'careers' and more casual and unrewarding 'jobs'. Exactly how this is manifest is a matter of organizational choice – the same technological configuration can be implemented with a workforce which is completely upgraded, or on the basis of polarization between career-track and more marginal workers – and may be influenced by public policies. Differences in the style and quality of vocational training and technology management across EC member states (and, we could add, across industrial sectors and firms of different sizes) are liable to mean significant differences in the trends displayed in specific instances.

This EC study also confronted the change that is taking place within IT itself – from stand-alone to networked systems, from computers and telecommunications to telematics and integrated solutions. Unfortunately, much of the empirical research covers only earlier generations of IT: it is important to examine its more recent evolution, since rather different implications may be associated with the new technological possibilities.

6.4.2 Business

New communications technologies allow firms to extend their contacts with employees and environments in two directions. Contacts can be *widened*, for example mobile communications mean that people can be reached at more times and places than heretofore. Contacts can be *deepened* with more extensive exchanges of information and more rapid updating of data, as is made possible by the extension of data communications networks within organizations, and the introduction of systems such as electronic data interchange (EDI) to mediate their transactions. We are beginning to see information that was previously handled solely on paper – and even information that first enters the firm in a paper form – being captured, stored, circulated and manipulated and acted upon within networks by means of Document Image Processing systems.

These developments mean changes in the received wisdom as to how organizations best function, as the transaction costs for various communications are reduced. Firms are liable to be motivated to seize the inherent possibilities by the search for competitive advantage. Reduction of waste and inefficient working practices (e.g. time spent searching for members of staff who are not in their offices, or for documents located in cavernous filing systems); reduction of delays between data being received at one point in the organization and acted upon at other levels; increased effort to match the organization's output with market or other salient demands; increased effort to understand and intervene in changing environments; these are parts of the strategies that may be pursued.

Some of the implications can already be deduced from observation of use of existing IT capabilities within vanguard firms – for example, in the much-vaunted development of 'flexible specialization' (rapidly adjusting production to market signals, and thus often producing in much smaller batches than was traditional) and 'network firms' (with new decision-making structures and organization based more on alliance between a number of semi-autonomous units, linked by new types of contractual arrangement). Other implications are likely to become apparent in the next few years. Among the most important responses we would stress three in particular (Figure 6.4):

1. a flattening of organizational hierarchies, with a concomitant displacement of middle management positions;
2. a shift away from traditional models of centralized office work, with some development of telework and, perhaps more significantly, new types of mobile work (e.g., moving between clients), and decentralized office structures; and
3. associated with the above trends, new responsibilities placed upon workers at lower levels of the organization, with a concomitant upgrading of their decision-making skills – but also, perhaps, some increase in the stress which they experience in their working life.

The scope for change is profound. There are indications that the potentials are already being taken up by a few managers, but more generally there is limited awareness of their significance. Management training will be one of the factors determining the speed with which, and

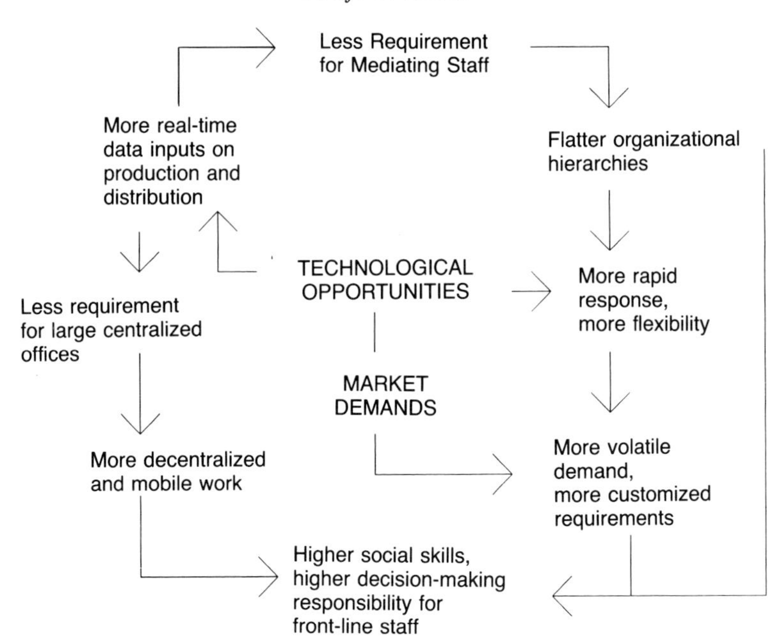

Figure 6.4 Change in work organization.

directions in which, such potentials are realized. Will the need for more responsible staff lead to more consultation with the workforce about change? Current levels of involvement of workers in decisions as to technological change are very low, although there is evidence that this is slowly changing, and that the experience is generally regarded positively by those managers and workers that have undertaken it (cf. the interesting surveys by the European Foundation (1991)).

Will job design be forced to accommodate to a greater extent to the social, psychological and economic needs of neglected parts of the workforce? These are questions whose answers are only partly determined by the cultures of specific companies, and where societal choices may be increasingly important as we enter the 21st century.

6.4.3 Transport

Physical transport has always been an important means of communication, with people travelling to exchange information and take joint decisions, and with goods moving in response to these decisions. While much has been written about the 'telecommunications-transport trade-off', it is notable that new communications facilities have so far tended to complement rather than substitute for transport. There is an interesting parallel here with the use of paper, which new IT has rarely reduced as was forecast: many innovations have actually boosted demand for paper as more documents, drafts and copies can be rapidly produced. (There is certainly scope for reductions in paper use in coming decades, as Document Image Processing and similar systems are built into office practices. But other 'paperless' innovations may actually increase paper use unless organizational change accompanies them. For example, EDI, if used so as to speed and increase the number of transactions, may boost the volume of conventional paper records generated as by-products.)

Emerging technologies such as computer-based videoconferencing (not to mention Virtual Reality!) may mark a change in this pattern. But it is equally possible that enhanced and intensified telecommunications will lead to more demand for face-to-face contact. The reason is that much of our business and personal life depends upon less formal interchanges, which tend to be suppressed during costly use of networks.

It is interesting to compare different types of communication link in this light: we might speculate, seemingly paradoxically, that asynchronous 'messaging'-type services – electronic mail, voice messaging, etc. – may foster more informality than conventional real-time contact, precisely because users are less likely to feel the clock ticking away as they interact. For an empirical comparison of new media, which shows that asynchronicity has several significantly different components for the user, see Lea (1991).

Likewise, organizational innovations associated with 'just-in-time' production and 'flexible specialization' may lead to increased frequency of shipments of (smaller batches of) goods, with increased demand for transport services (even if improved routeing, and optimal packing are introduced). Together with the advent of the European internal market, growing pressures of urban congestion, and environmental problems, transport is liable to be a critical area for innovation into the 21st century.

A recent study of these issues argues that IT will be critical to transport innovation, so that we will see communications systems as fundamental components of a strong transport informatics sector (Hepworth and Ducatel, 1992). It highlights the scope for the application of IT to 'pay-per' use of roads and other infrastructural facilities – a point relevant to areas other than transport too (broadcasting is the obvious, but not the sole, example). 'Wired highways', new information services and businesses, offering assistance with route planning and logistics, and innovation in emergency services, are on the horizon and in some cases already in evidence. But the precise course of change will be critically shaped by policies for public transport, land-use planning, and urban redevelopment. Rather than being marginalized by the IT revolution, the issue is really one of how transport will move into the information age.

6.4.4 Politics

It is in the sphere of politics that the perspectives of the different world-views of Figure 6.1 are their most divergent. Have the new technologies played a significant role in the liberation of Eastern Europe, or are they instruments for closer surveillance of us all? New IT permits use of communications *both* for citizen mobilization and creative social and political action, and for official (and corporate) monitoring of group and individual behaviour patterns; *both* for enhancing political awareness and understanding and for drowning reflective thought in information overload. For forecasting purposes, the safest assumption is that the normal checks and balances of Western democracies will continue to check extreme scenarios – but the safest assumption may still not be a really safe one. The more pessimistic outcomes in which IT is married to authoritarianism are imaginable, and the seeds for such developments could be present in the rise of extremist political movements (e.g., the National Front in France), pressures for tougher responses to pervasive violent crime (especially if tied to the rise of an underclass), and moral panics concerning disease or socially deviant behaviour.

Clearly there is a strong imperative for developing institutions and ethical norms which can inhibit any such drift in the use of IT. Data protection law is a step in this direction, but a limited one. There has been little debate – even among experts – of the far-reaching possibilities for using new communications networks and transactional systems to track individual activity; yet intelligent networks and new financial services implicitly demand the creation of new data bases of vast scope. In the long run, these are far more salient issues than the minutiae of telephone charging, which seems to be the primary topic of concern about telecommunications at present.

Some of the more visionary prophets of new IT have argued that vastly enriched political participation is liable to be brought about by the application of IT to sampling citizen opinion, facilitating virtual meetings in which constituencies can be created and decisions made, informing people about parliamentary and other proceedings which may affect them, and so on. To date efforts to institute 'teledemocracy' have proved rather disappointing; see the review by Arteton (1987). We doubt that enhanced and sustained involvement in politics on any scale is at all likely without, at the very least, substantial changes in our political institutions. What is much more likely, however, is a gradual extension of such innovations as the use of electronic mail as a means for representatives to keep in touch with their electorates (used by a few pioneers in the USA), bulletin boards as means for pressure groups to liaise and rapidly disseminate information (as various 'green' groups did during the Chernobyl disaster), computer conferencing as a forum for debate over policy (e.g., Bitnet conferences on issues of computer crime and ethics, and on the Gulf war), and social experiments such as the 'Electronic Village Halls' currently being introduced in Manchester in an effort to bring marginalized groups and areas into the information age.

6.4.5 Culture

For many commentators, the coming of electronic communications was bound to lead to a 'massification' of culture. The same ideas, artworks, and media personalities would be disseminated throughout the world, forming a common set of cultural referents. In pessimistic versions of this vision, a lowest common denominator mass culture would be pervasive – 'wall-to-wall Dallas' broadcast from satellite channels, with apparent proliferation of choice masking poverty of content and an erosion of high culture. While it is undeniable that there are elements of a global media culture, it is less evident that this is resulting in an homogenization of people's attitudes, values and tastes. Indeed, already in the 1970s survey research was being used to refute the notion that media, education and social mobility were promoting massification – opinions do converge on some issues, but equally they diverge on many others; see Glenn (1974). More recently, a series of studies for the EC's FAST programme have demonstrated that there is considerable diversity in the application of technologies in European ways of life – and have argued that this diversity is actually a source of strength, providing a fertile ground for the evolution of innovations. For a summary of the numerous studies here, see Hingel (1992).

More recent commentators have enthused about the emergence of 'postmodern society', though there has been both criticism of the concept and of the celebration that often goes along with it. These commentators have argued that we are witnessing a break-up of the project of modernity, with its assumptions about an unfolding historical narrative leading to a more rational society (implicitly organized along traditional Western lines). Part of the argument about postmodernity suggests that we are moving into a world in which we are bombarded by media images, and communicating with people from diverse cultures and subcultures: we are thus subject to an almost continuous kaleidoscope of social perspectives and of fragments of culture which remain only weakly attached to their origins. In place of the 'modern man' of the 1960s, commentators talk about 'post-modern people' with multiple selves, informal behaviour, and aestheticized lives, pursuing fashions and unwilling to commit themselves deeply in personal relationships of political ideologies. For very different perspectives on these issues, see for example, Barns (1991), Gergen (1991), Featherstone (1991).

The emergence of post-modern identity, while a much-debated issue, is central to any detailed discussion of the social implications of emerging technoeconomic formation involving advanced communications facilities. Despite hyperbole and casual use of empirical evidence, we do see here an effort to address questions of our psychological and cultural responses to the explosion of communications opportunities. Yet it would be unwise to assume that trends to post-modern diversity are themselves homogeneous. Some – perhaps many – social groups may well react defensively to the threats that pluralistic societies may be rightly or wrongly be seen as posing to cherished values and practices, and rather than reducing their effect, they may revert to fundamentalism or neotraditionalism of various kinds. (This suggestion was already being made some twenty years ago in discussions of 'collective search for identity' (Klapp, 1969) and 'future shock' (Toffler, 1970).)

6.5 CONCLUSIONS

The information age means a transformation of agendas in all policy areas. It is no surprise that most politicians and civil servants are preoccupied with the short-term, and that many academic researchers are devoted to the narrow problems defined by their disciplines. We lack even basic statistics on many features of IT development, diffusion and use (cf. Miles *et al.*, 1990). There needs to be more opportunity for serious consideration to be given to the sorts of issue that have been discussed above, with more extensive and systematic research. The future is too important to be left to the futurists.

It is interesting to note that several countries have state agencies which are devoted to the analysis of long-term issues, and that interesting studies on communications futures have been produced by our transatlantic cousins in recent years; see Department of Communications (1987), NTIA (1988), OTA (1989). The most substantial report, by the US Office of Technology Assessment (OTA), demonstrates the value of such an institution. In Britain the notion of technology assessment has never fully taken off. It is currently being strongly promoted in Europe, but the European Parliament's OTA-equivalent, STOA, does not seem to be focused on long-term communications issues, and the EC's FAST programme, after a period of interest in 'information society', has set itself other priorities. Change may be forthcoming as a result of the

European Parliament's insistence that technology assessment and evaluation accompany all major R and D programmes. But at present the UK, and Europe, lack institutions to supply us with the sort of foresight needed for wider debate on what sort of information society we want to achieve. In the end, the social implications of communications in the 21st century will be the result of our actions and choices. Can we afford to perpetuate a situation where only compartmentalized experts debate the profound implications of change, with most of the public and even the political elite remaining poorly informed?

REFERENCES

Arteton, F.C. (1987) *Teledemocracy* Sage, Newbury Park, Cal.

Barns, I. (1991) Post-Fordist people? *Futures* **23**, (9), 895–914.

Baum, D. and Graham-Jones, S. (1989) *Child Health: the Complete Guide*, Penguin Viking, London.

Beniger, J.R. (1986) *The Control Revolution*, Harvard University Press, Cambridge, Mass.

Department of Communications, (1987) *Communications for the Twenty-First Century*, Government of Canada Department of Communications, Ottawa.

Dosi, G., Freeman, C., Nelson, R. *et al.* (eds) (1988) *Technical Change and Economic Theory*, Pinter, London.

Ducatel, K. and Miles, I. (1991) New information technologies and working conditions in the European community *Information and Communication Technologies: Social Aspects, Employment and Training*, EEC Conference 17–18 October.

European Foundation (1991) *Participation in Change*, European Foundation for the Improvement of Living and Working Conditions, Dublin.

Featherstone, M. (1991) *Consumer Culture and Postmodernism*, Sage, London.

Freeman, C. and Soete, L. (1991) Macro-economic and sectoral analysis of future employment and training perspectives in the new information technologies in the European Community: Executive summary, policy conclusions and recommendations, mimeo, MERIT, University of Limburg; presented at *Information and Communication Technologies: Social Aspects, Employment and Training*, EEC conference 17–18 October 1991.

Gergen, K.J. (1991) *The Saturated Self*, Basic Books, New York.

Glenn, N.D. (1974) Recent trends in intercategory differences in attitudes, *Social Forces*, **52**, 395–401.

Hepworth, M. and Ducatel, K. (1992) *Wheels and Wires*, Pinter, London.

Hingel, A. (1992) *Science, Technology and Social and Economic Cohesion in the Community – a long term analysis. Overall synthesis report* FAST Prospective Dossier no 2, Vol 1, FOP 300, European Commission, Brussels/Luxembourg.

Klapp, O.E. (1969) *Collective Search for Identity*, Holt, Rinehart & Winston, New York.

Lea, M. (1991) Rationalist assumptions in cross-media comparisons of computer-mediated communications, *Behaviour and Information technology*, **10**, (2), 153–172.

Miles, I. (1988) *Information Technology & Information Society: Options for the*

Future, Economic & Social Research Council, PICT Policy Research Papers No. 2, London.

Miles, I. and Robins, K. (1992) Making sense of Information, in *Understanding Information* (ed. K. Robins), Pinter, London.

Miles, I., Rush, H., Turner, K. and Bessant, J. (1988) *Information Horizons: the long-term social implications of new information technology*, Edward Elgar, Aldershot.

Miles, I., Davies, A., Haddon, L. *et al.* (1990) *Mapping and Measuring the Information Economy*, British Library (LIR Report 77), Wetherby: Boston Spa.

NTIA (1988) *NTIA Telecom 2000*, US Department of Commerce, National Telecommunications and Information Administration, Washington DC.

OTA (1989) *Critical Connections*, Congress of the United States, Office of Technology Assessment, Washington DC.

PACE (1992) *Perspectives for Advanced Communications in Europe*, European Commission, DG12, Brussels.

Rush, H., Miles, I., Bessant, J. and Guy, K. (1986), *IT Futures Surveyed*, NEDO, London.

Ryan, G. and Aboulafia, A. (1992) *ESPRIT Project 5374: QLIS: ED 2.1 First Version of the Generic Reference Framework*, Copenhagen Business School (mimeo), Copenhagen.

Toffler, A. (1970) *Future Shock*, Bodley Head, London.

7

The impact of networks on financial services: the reality of electronic banking

J. De Feo, Chief Executive, Information Technology and Service Businesses Division, Barclays Bank PLC

7.1 INTRODUCTION

In the last 15 to 20 years we have seen a massive development in the usage of computing technology to capture and process information, both to speed up the work process and to enhance information analysis.

Although the resultant change in working practices has been significant, particularly in manufacturing industries, many of the promised benefits of computerization, in particular, reduced costs, greater efficiency and competitive advantage derived from the timely leverage of information, have not been optimized. Whilst raw computing capacity to underpin the benefits is now available, its exploitation has been restricted by the slower evolution and relative expense of the telecommunications and networks capacity compared with computing power.

However, the 1990s will see a major change in both the economics and the flexibility of the telecommunications infrastructure, such that the type of developments that were previously the realm of science fiction will become reality, and could revolutionize the way we live and work. We will then finally see at the turn of the century a move from the second wave 'industrial age' to the third wave 'information age', which has been forecast now for many years.

This paper will seek to:

- outline some of the developments we might see in 'raw' computing and telecommunications in the next 5–10 years;
- describe some of their potential uses;

Communications After AD*2000*. Edited by D.E.N. Davies, C. Hilsum and A.W. Rudge. Published in 1993 by Chapman & Hall, London, for The Royal Society. ISBN 0 412 49550 3

- outline the way they will impact our lives;
- assess the types of changes we will see in the demand for and supply of financial services; and
- assess the resultant changes we will see in financial services in the 21st century, as we prepare for electronic banking in an information age.

7.2 TECHNOLOGY DEVELOPMENTS

7.2.1 Computing and processing

The fundamental enabling technologies of computing are continuing their steady incremental improvement of becoming better, faster, cheaper and smaller. The power of room-size computers of the 1960s is now available on the PC at your desk. As the trend of miniaturization continues, the same amount of power could be available in your wrist watch before the turn of the century.

In raw terms, the operating speed of computers doubles each year, and is likely to continue this way for the next decade or so. Computing cost has been halved approximately every three years.

Parallel processing will provide enormous power with resilience and speed – Cray supercomputers will operate at 1 000 000 million instructions per second by 1995 – but more importantly for commercial situations, such power is scalable, with processing power added or removed in small increments to match demand.

Coupled with the above is a commensurate development in storage capacity and cost reduction. Storage costs on magnetic media have reduced so much that, if present trends continue, storage costs will no longer be an issue in computing of the future.

7.2.2 Communications

Like computing, communications technology has grown exponentially. However, the actual usage has not reflected this level of growth, principally because of the inadequate speed of current electrical switching technology. However, with the increasing use of optical fibre cabling and switching, the development of integrated broadband networks, and the growth of satellite and mobile communications (radio and infra-red), this picture will change dramatically.

As the cost per bit of optical transmission continues to fall, the need for video, speech and data compression currently required for this mode will ultimately disappear. Bandwidth efficiency will no longer be an issue; users will be more interested in widespread availability, flexibility, reliability and price.

The use of optical fibre is already widespread, servicing over 55% of all international traffic, and over 80% of all UK and 65% of all US long distance national calls. However, its full potential is limited until it can be made available in the local loop (i.e., to the home). Numerous experiments are underway to achieve this.

In Europe, trials are underway in the UK, Germany and Spain, whilst France already has a 'Mission Cable', designed to install optical fibre networks around most French cities. In the UK, BT are experimenting in Bishop's Stortford with TPON (telephony passive optical network), carrying telephone channels and low-speed data, and BPON (business passive optical network), capable of carrying telephone, TV signals, hi-fi radio and a video library. In Germany, field trials by DBP Telekom and Raynet are underway using PONS for telephony and Cable TV, and similarly in Spain. These trials are only a foretaste of the ultimate possibilities, and will probably be extended significantly during the next few years. The likely take-up of PONS in Europe over the next decade will be significant with Germany and the UK leading the way.

Wider installation of ISDN during this decade will be a major step forward in the potential for transmitting non-voice communication, albeit compression techniques will still be required for some non-voice media. However, the full potential for networks communications will be realized when true broadband networks are available, providing 'bandwidth on demand', and thus enabling multimedia transmission of information wherever and whenever it is required.

Fibre-based communications will be complemented by and will be in competition with satellite and radio communications which will create a fully global telecommunications infrastructure. The compatibility of the various elements within this infrastructure will need to be determined over the next decade by the rigorous and timely setting of standards. Although the capability of the infrastructure will be technically inexhaustible, its actual flexibility will be determined by the foresight of the infrastructure providers and users in specifying their needs.

Once 'bandwidth on demand' is available, the real potential of computing power and flexibility can be unleashed. The next section looks and some of the likely uses and developments of computing in the next 5–10 years.

7.3 THE TECHNOLOGY POTENTIAL

7.3.1 Distribution of information and processing

Technologies already exist to separate the provision of computing power from where it is used. These include client server architecture, and the use of distributed database management systems, such that the user may access information at his own workstation, regardless of where the information is stored or processed. The user's workstation may be on a LAN, or alternatively on a wider communications structure such as a metropolitan area network (MAN) or a wide area network (WAN).

Today there are significant performance implications affecting the distribution of information. Whilst the use of a LAN can provide timely information transfer within a restricted area, the use of a MAN or WAN is generally slower, restricted by the need for standards, protocols and adherence to regulatory requirements. In future, once standards have been agreed, the location of the information user, the information itself, and the means of communication between them, will be transparent, as the network infrastructure seamlessly handles the transmission of data to and from remote storage and processing centres.

7.3.2 Handling new types of information

It is already possible to handle multimedia communication through creation of compound documents. New techniques under development include electronic image, voice recognition and generation, which extend computing into pictures and speech. Once broadband networks are available, this type of information can be transported without impairment wherever required.

The simulation capabilities of computing are already extensively used in CAD/CAM applications. However the concept of virtual reality will develop strongly, with moving holograms used not just for business or engineering, but also for entertainment.

7.3.3 Extended scope of computer processing

As outlined in section 7.2.1, the trend towards miniaturization of computing will continue, and we will see increased use of smart cards and pocket (or watch) PCs. The widespread use of the latter with mobile communications could revolutionize the efficiency of mobile sales forces.

Advances in artificial intelligence (AI) will alter the way in which computers are used. Knowledge based systems and neural networks will become commonplace in the next century, operating on local and international levels.

7.3.4 Software developments

Software is a relatively new science in computing, and to date has not developed as fast as the hardware technologies. However the pace of change is quickening, as users move from processing to leveraging information. Software is becoming more flexible and user-friendly, in order to meet the demands of increasingly sophisticated users for information manipulation and analysis. Relational database management systems are commonplace, and will be superseded by distributed database management systems, once the telecomms capacity for handling such systems on a timely basis is in place. Further flexibility in software development is promised by object orientation, which will allow the re-use anywhere on a network of tried-and-tested software elements to meet common needs.

7.3.5 Low-cost storage

The development of low-cost optical storage disks, together with image scanning techniques, makes the recording of information and subsequent access to it a simple task. Combined with data compression techniques and fast transmission, the storage and ease of access to huge libraries of information on an international basis has major implications for education and commerce.

7.4 LIFE AND WORK IN THE INFORMATION AGE

How the new and emerging technologies outlined in section 7.3 will impact the way we live and work will depend on the speed at which we are able to absorb the fundamental changes they could bring to our lifestyle. This section will suggest some of the changes that could occur, some of them beneficial, but some fraught with risks.

7.4.1 Domestic life

Sometime in the 21st century homes will be on-line, with access to all available information sources. The homes themselves will be intelligent, with systems capable of adjusting a room setting to suit the people in it – e.g., reading lights or TV screen available for entertainment, or systems to control the security of the home. A key feature will be the invisibility of technology.

Homes will be equipped with high definition TV (HDTV) flat screens which will be the device for access to and receipt of all visual information. The device will be activated by voice input, though there may also be a pen or keyboard or mouse, dependent on the preference of the user. Screen-based personal jotter pads will also be available for use around the home e.g., an electronic shopping list, with orders transmitted at a given time direct to the supermarket.

Security and privacy are issues of concern to technology users today.

Next century they will still be an issue, but technology will be used to provide such security, for example the use of super smartcards, with inbuilt voice prints and signature recognition which holders can use to identify themselves in the following ways:

- to a terminal device, which is then set-up to the user's liking;
- for access to privileged information or service – e.g., membership of a particular club, library, etc.; and
- for financial transactions.

7.4.2 Social interaction

Man is inherently a social creature, so it is unlikely that technology will be used to such an extent as to isolate him from physical contact with others. There will still be demand for leisure centres, high street shopping, doctors, dentists, hospitals, pubs, etc. However, at each such site there is likely to be the wherewithal to access information electronically. For example there will be information booths in high streets to allow customers to browse through shopping catalogues/brochures, then selecting an item for physical examination before purchasing, or to pay bills, or schedule/cancel appointments etc.

It will also be possible to socialize electronically with others by means of video telephones, electronic mail, bulletin boards etc. This has already started, for example in the use of the mail element of the U.S. based 'Prodigy' service. Whole electronic communities or 'tribes' may well develop.

7.4.3 Leisure/entertainment

The entertainment industry is likely to be a prime initiator of new products and services in the flexible networks environment. Computer games already have a strong foothold in the market – for example Nintendo games are found in more than 70% of all US homes with a child between the ages of 8 and 12. The capacity to network these games might on the one hand add a new competitive dimension, but on the other could run up extortionate communications bills, not dissimilar to the infamous telephone chat-lines.

It will be possible to select electronically films for home viewing, reviewing trailers of the films on screen to help you make your choice. It would also be possible to create your own film by selecting the characters and the outcome you wish to see. The enabling technology (the compact-disc video) is already under development.

Simulation, using holographic video or virtual reality, could prove extremely valuable if used in an educational way – e.g., in helping to understand how things work. However, there is a real risk that virtual reality could turn out to be the LSD of the 21st century, with users so 'hooked' that they are unable to distinguish it from the real world.

Advertizing will be revolutionized – the Yellow Pages directory would become real-time and on-line, such that you could make direct enquiries on an advertiser's price list, and see the goods before placing an electronic order. Holiday brochures could be distributed on CD, with users specifying their requirements, and the brochures displaying only those holidays which match the users' criteria.

The concept of a personalized magazine or newspaper could become a reality, with news being displayed on only those items which interest you.

7.4.4 Education

Computers are already playing a vital role in all levels of education. We rely on universities and research institutes to develop the leading edge of technology. However, we can also see technology becoming an integral and invaluable part of general education. In some schools in the US and in remoter parts of Australia, networks are already providing direct video links between teachers and pupils unable to attend conventional classrooms. Education for the handicapped could be transformed.

Computer simulation will allow pupils to understand far more graphically than before the structure, functions and relationships of the elements they study. Computer screens will be part of standard school desks, forming a jotting pad, where pen or voice input will activate required processes. Custom-tailored text books and teaching courses will also be developed for electronic transmission.

The electronic availability of vast libraries of information, either by CD-ROM or by on-line access will stretch students' ability to access and synthesize information. They will be helped by the use of software 'agents', semi-intelligent objects created in the computer as personal assistants, skilled at extracting information and customizing their availability to the student's needs. However, notwithstanding the power of such agents, educationalists must be aware that computers at best should be a stimulation and aid to learning, not a direct replacement of the thought, reasoning and synthesis process.

7.4.5 Working practices

By the 21st century, electronic interactions will be a common feature of doing business. Computer networks will profoundly affect the structure of organizations and the conduct of work.

At present face-to-face exchanges are still the predominant form of communications in organizations. Once the use of electronic communications has reached a critical mass, the need for such contact will diminish in many industries. Communication will be predominantly by means of electronic mail, video telephony and conferencing.

There will be less need to work from an office. Mobile communications will greatly enhance the capabilities and professional approach of mobile work forces. Telecommuting will become more prevalent, as home based computing and communications facilities will match those available in an office. This could ultimately lead to a migration away from cities or 'commuter-belt' areas.

7.4.6 Organization structure

Employees at all levels of the organization will have access to all types of information. They will be invited to contribute to organizational issues on the basis of their expertise and knowledge, rather than their level or grade. As a result, the current hierarchical tendencies in organization structures will disappear.

Employees with specialist skills will also be invited to form work groups brought together to solve specific strategic or operational issues. Such teams can be formed at any time, and it is possible for employees to be involved in several such work-groups at once, sometimes without the knowledge of their line managers. This will create pressures on current management practices, leading to significant changes in work control, measurement and reward etc.

The ultimate development here would be the formation of virtual companies, whereby, for example, a group of designers would bring together groups of individuals with specific expertise to assist in producing, distributing and selling their design idea. No legal 'company' as such would exist, but the partnership of talents would operate effectively together with groups dropping out of the partnership when their services were no longer required. Real-time communications mean that design processes that were previously undertaken consecutively can now be handled in parallel, with the result that design and production time scales are dramatically shortened. Product lifecycles are also shortened, as individual customer demand will call for continuous enhancement and change.

Such a model can only exist once world-wide telecommunications are sufficiently developed and standardized to form an invisible part of the operating infrastructure. Once in place, market economies would come to the fore, with buyers and sellers seeking each other out in the global information market-place. The implications are significant for all industries, but particularly so for financial services.

7.5 THE FINANCIAL SERVICES INDUSTRY

7.5.1 Introduction

The financial services industry has information as its prime raw material – information on its customers, its competitors, the national and global economy etc. The services it sells consist for the most part of information – on transactions across an account, on advice given in the case of mergers and acquisitions, on sales or purchases of stocks and shares, on assessments of future currency values (in foreign exchange contracts, swaps and options) etc.

Networks, as the means of access to and distribution of information are a key component of any financial services company's infrastructure. Today they are the work-horse, handling transmission of information processed from the mountains of paper handled throughout the industry. In the future, however, they can be better harnessed to add value by enhancing and delivering existing and new services electronically, and by reconfiguring the structure and representation of the industry as we know it.

7.5.2 The traditional role of financial institutions

Banks have traditionally offered lending and deposit-taking services, and have, by virtue of their licences, been able to create money by onlending. However this traditional franchise is open to challenge. Credit is already offered to personal customers by non-financial institutions or at point of sale, and in the corporate market, credit is provided by the investment market through the process of disintermediation. Deposit taking is no longer the preserve of banks, except for those customers who find it inconvenient to invest directly with the market.

Banks still have key control of payments mechanisms, and by virtue of their reputation for safety, integrity and security, handle virtually all payments traffic. However, with the further development of EDI, this may not remain the case, particularly if corporates were to develop their own clearing house for trade credit. Whilst the notion may sound extreme, it serves to underline that all areas of the banks' business could be under threat from new competitors who could, with the aid of technology, re-engineer processes and thus break down the barriers to entry that have historically preserved the banks' dominant position.

7.5.3 The challenge for the 21st century

The challenge now facing all financial institutions is to leverage technology ahead of the competition in order to:

- provide quality service for increasingly sophisticated customers; and
- run processing operations at the lowest possible cost consistent with the quality required.

Coupled with the above is the need for financial services companies to develop and manage the customer interface of the future. Electronic communication will become the norm, but not the only form of contact. Banks must therefore be able to develop and tailor products to suit the specific needs of their customers and capable of delivery by whatever means the customer requires.

Yet multiple delivery channels will not be enough, as most institutions will be able to offer them. The most successful companies in the next ten years will be those which are able to anticipate customer demand by envisioning products ahead of their time, and ensuring that they are ready for market when the technology and the demand arrive. As the pace of technological development is shortening the product lifecycle, the challenge and pressure to create new products will become greater – the ability to innovate will therefore be the critical differentiator of the future.

To preview some of the changes which may come about, we will look first at the likely requirements of personal customers, then at corporate requirements, and finally at the needs of multinational or global operations. (Throughout this section, the term 'banks' will be used to cover both banks and other financial services companies.)

7.5.4 Personal customers

Home banking will be the norm. Banks may supply software, which in addition to providing the customer with access to his own records on the banks' systems, and allowing for payments, transfers etc., also provides a vehicle for home shopping, holiday booking, appointment fixing, social diary etc. Wherever payment for services is required, such payments can automatically be initiated through the bank.

Alternatively a retail organization may provide a home shopping capability, with banks providing instantaneous credit rating and a payments mechanism to facilitate the purchase.

In both examples, real-time on-line information handling is required. Knowledge-based systems will be used to process the credit-rating and issue payment instructions.

From the banks' point of view it is clearly more beneficial to provide the home banking capability themselves, as it:

- opens up a key delivery channel;
- helps maintain a hold on the payments mechanism; and

- provides a vehicle for the sale of financial products – e.g., insurance, pensions, stocks/share, financial advice etc., where value can be added.

The implications for the banks of this new electronic delivery channel direct to the customer include:

- means of capturing more reliably information on customers' spending habits;
- ability to analyse, through a knowledge-based system (KBS) customers' spending habits and to develop/sell new products accordingly;
- unbundling of existing products and restructuring and re-pricing to meet needs of individual customers – this might include the re-packaging of wholesale/derivative products for the sophisticated personal customer;
- separation of delivery from processing, and the ability to drive down costs in both areas;
- credit control and sanction are no longer an issue – they are instantaneous decisions and can be handled by a KBS; and
- more time and resource available to be proactive sellers/designers of new products.

Banks will still need to maintain a facility for face-to face contact, with their customers, particularly for advice, a financial planning service, and sale of complex financial products. This can be provided by:

- video telephony,
- mobile sales forces, with home calls if required; and
- mobile sales offices, 'drop-in' offices at key events, or moving locations in line with sales campaigns or customer demand.

This will represent a fundamental shift from the current position of infrastructure-intensive local branch sites which are expensive to run. Processing will be moved to remote locations where economies of scale can be achieved. Local representatives will have immediate access to information via mobile, public or private (bank) networks, or indeed a mixture of the three.

The implications for the customer are also fundamental. He could have electronic access to a number of banks and could in theory send out invitations to several banks to tender for his business. He will be able to choose a product tailor-made to his requirements, and the timescale for his decision will be greatly reduced from today's norm.

Privacy and protection issues will be a major area for the banks to address: technology will have to be harnessed quickly and efficiently if banks are to retain their competitive strengths of integrity and security in the new distribution channels.

7.5.5 Corporates

A hint of the impact of the information age on the business cycle of corporates has been seen in the development of EDI. However, the capacity for further development here is enormous; the concept of an electronic just-in-time (JIT) business cycle from raw materials through to finished product could be a reality early in the next century.

A customer will initiate an electronic order for specific goods. This will trigger an instruction back down the entire value chain, such that, for some products at least, production can be tailor-made to clients specifications. This feature is already available in certain industries, (e.g., quality cars, haute couture), but could become more prevalent in mass-production industries, once the communications infrastructure is available. The key issue for banks is that, at each point in the value chain, (e.g., the sale of raw material to the producer) transfer of value takes place. Banks must be represented at each point if they wish to preserve their settlement business. A full extension of EDI could achieve this, with banks acting as facilitators or as a clearing house for payments, and for associated record-keeping, reconciliation, netting etc.

Lending to firms for working capital may be less prevalent than before, partly because of capital constraints and requirements on banks, and partly because the EDI cycle outlined above will help balance the supply and demand for trade credit among the firms themselves.

Funds for capital investments may be provided by banks (with credit decisions automated), but will more likely be provided by other investors. The banks' role will be to act as agent, e.g., in putting together a workgroup to structure a proposition, seek investors, give advice etc. This is a similar role to that undertaken today in syndications, but it will be carried out electronically, and will be enhanced by the speed of information-gathering, assimilation, analysis and decision-making. Experts will be on hand (electronically) for tax/investment advice etc., and the customers will participate electronically throughout the process.

Product development will speed up, as in the case of personal customers. Tailor-made solutions will be the norm.

The implications for the bank include:

- virtual absence of physical presence in an office;
- use of mobile offices or direct visits to customers if face-to-face contact is required;
- ability to develop products in reduced timescales;
- the need for large-scale efficient processing of payments, records etc.; and
- the need for fast, reliable networks to allow customers to access their own information, and for banks to keep control of the value side of the EDI cycle.

7.5.6 Global institutions

The traditional bank role of lending and deposit taking has already virtually disappeared in the global market place. Banks' current role as agent/broker will expand to become the norm, particularly as corporations themselves will have equal or greater knowledge than the banks on the investment opportunities available in the market place.

In the treasury arena, imperfections in the market arise traditionally through differences in perception of market movements, of events or in timing and quality of information available. Improvements in networking technology will help drive out imperfections in the timing and quality of information available, such that it will be increasingly difficult to make money out of a position. This could significantly alter:

- the concept of risk in the market place;
- the number and type of risk-related products available;
- the ability to make money on speculative trading deals; and
- the number of players in the global market.

Some banks will still be in the business of trading, not only to satisfy their own and customers' needs, but also to help maintain liquidity in the market. However, real-time access on a truly global basis to information, and artificial intelligence applications to help act on the information is a fundamental prerequisite.

The banks' risk-related skills will still be required, for example for underwriting debt issues, though perhaps to a lesser extent, as the time lag and hence market risk will be reduced.

Banks can exploit the global market place by enhancing their cost effective mass processing of information on a global scale. This is already taking place in securities handling, though it can be improved by real-time information, and by dematerialization in all world-wide markets, both of which require fundamental improvements in information processing and transmission.

Processing capability can also be sold in bulk in other areas, for example a global payments service, computer operations service and use of banks' networks by third parties. The networks themselves will have become a strategic corporate asset, to be exploited both inside and outside the financial services industry.

The key requirement in all these areas is to harness technology to provide processing on the cheapest basis in the cheapest locations, using the networks to move huge amounts of information on demand to wherever required, and to allow access to it by a wide range of legitimate parties.

7.6 ORGANIZATIONAL IMPACT

From the above it is clear that the financial services industry as we know it today, and the people working in it will need to undergo major transformation in order to meet the needs of customers in the information age. The transformation will encompass our operations, personnel requirements, organization structure, and corporate culture.

Operationally, our current strengths in paper-based processing skills will need to be transformed into electronically based processing centres of excellence. Our products will need to be capable of electronic delivery and access. Our networks will need to facilitate transmission of bank and non-bank information, in a manner and at a speed which suits the customer, and not just the bank. Our applications will need to be flexible enough to allow a wide variety of tailoring and development of products, and to allow direct access by customers at will. Our whole infrastructure must become flexible and mobile, to allow us to accommodate customer requirements wherever and whenever required.

As far as the skills of our personnel are concerned, our traditional strengths of credit risk assessment will be taught to computers, whilst we will need to develop an understanding of new risk concepts, that have not previously been the realm of banking. Our clerical staff will no longer be processors, but professional sales and service people, equipped with tools to help in predicting and determining customer demand. Our branch managers will no longer manage outlets, but will become operations managers and coordinators. All our staff will be both technology and business literate, working with both networks and computing suppliers and our customers to identify new services.

Our hierarchical structure will yield in favour of multidisciplinary project teams, and our senior management will have devised new ways of co-ordinating and measuring our performance based on our effective use and leverage of the information at our disposal.

Many of these trends are already beginning, but they will gain impetus in the next few years, and must be the norm by the year 2000 if we are to compete successfully.

A major question arises as to the ability of any of the existing financial institutions to achieve the above transformation. It is likely that only a few institutions will have the ability or the desire to remain active in all the areas covered in section 7.5. Most banks will become niche players with their own specialities, forging alliances and joint ventures with others to cover the rest of the market.

Regulation in financial services will be a major factor in determining the pace of change in the industry. The banks must be able to meet customer demand if they are to survive, and the regulators must therefore be flexible, whilst at the same time assuring stability and safety in an increasingly active global market-place. However, it is vital that non-bank competitors who have the capacity through technology leverage to provide financial services should be subject to the same regulatory treatment.

One thing is clear – no one institution can afford in financial or management terms the networks infrastructure required to run the financial services conglomerate of the future. Partnerships with corporations in the same and other industries and with networks and channel providers, both public and private, will be essential.

8

Impact of regulations on communications

Sir Bryan Carsberg, Director General, Oftel

I am delighted to be presenting a paper to the Royal Society since I had always hoped that one day regulation would be recognized as a science. Perhaps this is a good step in the right direction. The invitation to comment on the world after the year 2000, (even though now it is less than ten years away) would often be considered dangerous by somebody in my position. But on reflection I found that I could not actually resist the temptation. After all I shall be moving from Oftel in June, and the way the telecoms industry evolves after the year 2000 will be influenced a good deal by my successor. The predictions I may make, may therefore not come back to plague me, so much as somebody else.

Of course, it is necessary to begin by asserting that the basic trend in telecommunications towards liberalization will continue. It seems to me that it has been a clear success so far, that the impact of competition in encouraging more innovative services and better efficiency of operation is clear to see for everyone who cares to look. And I find, that a great majority of those in the industry and users also seem to share that view.

As one gets towards the end of what is almost an eight year stint at Oftel its fun to look back. And looking back is quite a shock actually. I remember in 1984, when I was appointed to this job, a number of people told me that they thought I was crazy to think that I could get competition going in a serious way in telecommunications. But now it is widely accepted that at least some competition is desirable and it is very hard to find a country in the world which is not interested in some measure of competition. The question now has become not whether there should be some measure of competition or not, but how far it should be taken, and how best to bring it about.

Communications After AD2000. Edited by D.E.N. Davies, C. Hilsum and A.W. Rudge. Published in 1993 by Chapman & Hall, London, for The Royal Society. ISBN 0 412 49550 3

In the area of value added and data services and apparatus supply, for example, liberalization is a long way towards completion in Britain and in the countries of Western Europe. At least when the EEC directive is fully implemented in the fairly near future, there will be almost complete liberalization of value added and data services and of apparatus supply. I suppose it is quite likely, that by the year 2000, more and more countries will have come to accept that simple resale should be allowed along side provision of value added and data services.

Now for those who are not telecoms industry experts I should perhaps just explain that value added and data services are provided by using a network that somebody else actually operates. The distinction that is currently made in many places is to say that those networks can be used by others freely to provide value added services, which are enhanced services compared to ordinary telephone services. But the ordinary basic network service of telephone calls and ordinary data transmission is still the reserve of the network operator. You cannot use the network freely for basic telephony on a resale basis.

What I am saying is that the distinction between value added services and the resale of network facilities for basic services will eventually disappear. It has already disappeared in Britain. Here you are allowed to set up business using somebody else's network to provide basic telephone services in Britain without any limitation domestically. That came about by the decision of the Secretary of State acting on my advice. I had come to feel that allowing simple resale completely would not do any harm, and perhaps more importantly, that to distinguish basic services from value added services in this business, just did not make sense.

We were dealing with an arbitrary dividing line, and one should always be suspicious of arbitrary dividing lines in my view. The ability to distinguish between many value added services, which involve a data transmission and basic voice services in an increasingly digital environment was, in practice, questionable anyway. So for that kind of reason the UK decided not to maintain the distinction, and I predict that allowing competition in value added and data services in other European countries will turn out to be the Trojan horse for allowing competition in basic telephone services. I think in time, they also will discover that it is not a sensible place to draw the line and I think that will be increasingly true, in other countries of the world.

The question then will become should network competition, that is competition in the operation of networks, be allowed. That is the fundamentally important decision. I also think that competition in the operation of networks will be increasing, and I note the existence of another Trojan horse in Western Europe for this purpose. Many countries have allowed competition in mobile services already, in the operation of mobile networks that is, while they have not allowed that competition in the operation of fixed networks. It seems to me, as time goes by, that more and more mobile network services will come to be seen as in direct competition with fixed services. People will therefore find that they have competition in basic voice services through mobile, even though they did not really intend it as such, because they regarded it as a kind of premium service when they made the decision. At that stage, perhaps the point of delaying further competition on that side will seem a small one. I believe that these trends will continue, and I make no secret of the fact, that I think that's the right thing to happen.

I also believe that social goals of telephone networks will continue to be met, despite the fact that many people express the concern that competition makes it difficult or impossible to meet social goals. The particular goal, for example, could be the provision of a universal telephone service, to those in rural areas as well as in urban areas at much the same sort of price.

Now the reason people are concerned about that, is that they think competitors will come in and target the urban areas which are the more profitable. This, they reason, will force the incumbent to lower prices in the urban areas to meet the competitive effort and leave them with no alternative but to put up prices in the rural areas in order to balance their cost equation overall. In other words, competition will have eliminated the cross-subsidy from urban to rural areas that takes place at the moment. That concern I have heard expressed by representatives of governments from other countries where they have said, we cannot have competition because we attach importance to social goals. But of course we attach importance to social goals here, BT is continuing to meet them admirably and more effectively than it did before it had to face competition. Most importantly, one could always adopt regulatory means to make sure that these social goals are met, and if necessary one can raise a levy on the competitors to help pay for meeting them. If, for example, different competitors pay a levy of an equal amount, perhaps through interconnection charges, or perhaps in some other way, on certain classes of traffic, then you have secured even competition for that traffic and the levy can be used to finance the provision of services which are the subject of the social goal.

So I think its false to argue that competition gets in the way of meeting social objectives. We have not yet in this country had to activate the arrangements for a levy for financing these services because BT's advantage, in being initially the monopoly operator has seen it through so far with an ample ability to meet the costs itself. But we have provision in our regulatory arrangements to do something like that, if it becomes necessary.

I think the questions of the future about universal service actually will move on to a new set of topics. I think, for example, the question will be how we can use the telephone network more effectively to serve the needs of disabled people. It is no secret that I am very interested in the idea that profoundly deaf people should be able to have telephone service as a matter of right and that that right should be recognized as much the same as the universal service obligation for people in rural areas.

I am not sure if telephone service for profoundly deaf people is quite the right expression, actually, perhaps there is a word other than phone that one should use but, of course, it involves using text terminals and a relay to speak the message on to hearing people. Quite possibly, the desirability of securing universal mobile service will be much in peoples' minds after the year 2000. It seems very likely that by then the mobile networks will have developed to the point where they will be within sensible sight of that kind of a goal.

I should mention also, that the balance of prices is a matter of social obligation. And that the balance of prices is related sometimes to concern about the effect of competition and the effects of allowing simple resale. The problem, at the moment, is that call charges are higher, and standing charges are lower, than they should be, economically. They were that way to quite an extreme extent in 1984 and their being so is a result of political pressures on the industry over many preceding years. Standing charges fall most heavily on residential customers, and that's where the votes are.

Since then we have recognized the need for, or the desirability of, adjusting the balance. However technology has been changing costs with such rapidity that as fast we have run to change the balance, accepting there should not be too large a change in any year, technology has moved ahead and the imbalance continues to persist.

However, I would predict that rebalancing will continue. I make no prediction about the rate of it, however, because that is the subject of an ongoing investigation. I don't actually think that it is efficient to keep an imbalance in prices, because it tends to mean that the network is not used efficiently. People are excessively discouraged from using the telephone once they have it, by high call charges; that is an economic loss which we can remove. It is not necessary to redress the balance completely, since through having competition one could use financial arrangements to prevent a complete balance if one really wanted to.

In the area of competition in the operation of networks, here in the UK in particular, I predict with some confidence we shall see progress. After an initial period in which we did restrict competition in the operation of fixed networks to a duopoly, we decided about a year ago to open things up more widely so that there is now no fixed numerical limit on the number of competitors in the UK. It is now quite clear that the actual number will increase, provided present policies are maintained.

What is interesting in this area is that local competition, competition at the local level, I predict, will become widely established. This has been an important area of UK policy leadership; no other countries in the world I think, until just recently, have really stated the desirability of local level competition as an important goal. Some of them have assumed that the economics of the industry work against it; that you can have long-distance competition but you can't have local competition. I do notice a change in that emphasis now and I noted others becoming progressively interested in it. I see, for example, in the US some of the regulatory authorities are now starting to talk about, and plan for, future competition at the local level, for residential customers. I am not talking just about business customers, but for residential customers also, it will become a reality.

We see encouraging signs, encouraging from a regulatory point of view at least in the UK, in that today cable television companies are providing about 35 000 exchange lines to customers locally. That, of course, is a very small number in the total context of the UK, but it is increasing by about 4000 a month at present and it is a token that the cable television industry is now starting to develop as a serious business. In my view at least, one no longer need have serious doubts about its coming to viability.

But on the subject of local level competition versus long distance competition, if you ask what makes a particular part of the market a good one for competition you can develop a sort of list of qualities. An area where you have direct contact with the customer is a promising thing, where you actually provide service and are in a position directly to understand customers wants for innovative services and provide new packages of service. Indeed the packaging through tariff is also an important point there. The ability to differentiate products through advertizing is another part of it. All of those things are present at the local level, in greater degree than they are at the long-distance level in telecommunications.

Furthermore, the economics of the situation are by no means unfavourable in the long-term to local competition. Economies of scale, duplication in resources as a result of competition, tend to inhibit it, but any financial analysis tells you that there is little difference on the whole between local services and long distance, on that score. There are some local areas which are promising for competition on that basis and some which are unpromising, but the same is true of the long-distance level.

Furthermore, it helps a new entrant to be entering the market where established operators are relatively inefficient, and without making any comment at all on the overall efficiency of BT, I would say, with some confidence, that I believe them to be more efficient at the long-distance level than they are at the local level. Now, of course, the thing that has encouraged people initially to go in to long-distance competition first, is that prices are out of balance, the very point that I mentioned earlier. Long-distance call prices are higher than economic so this is a promising target for competitors to enter. There is a good ceiling, they think, for them to work to; but that is likely to be a transitory phenomenon.

Of course it can partly be made up by the kind of regulatory arrangements involving the pricing of interconnection, and we have now made it clear that that will happen sooner or later anyway. Furthermore, as I have also indicated rebalancing of prices is likely to continue and that also will alter the margins that are available to new entrants. So it would not be unduly rash of me to predict that we shall see a relatively high level of growth in local level competition in telecommunications, and that local level competition will become firmly established by the year 2000. Indeed, I strongly hope so.

On the subject of network competition, over the last year the review of competition policy has prompted people frequently to raise the question of the desirability of breaking up BT. I will say that this is not a cause I advocate and so I would not want to predict that it will have happened by the year 2000. I do however think that there is some degree of inevitability about accounting separation within BT; the growing level of competition, and the regulatory arrangements that are needed to support it, will make it inevitable sooner or later. BT would then have to establish a separate accounting division for local operations and long-distance and international operations, and operate from one division to the others through open contracts, the terms of which are transparent and equally available to BT's competitors both at the local level and the long-distance level.

It is has been convenient for all of us not to bring that situation about in the early years and in my view the time is perhaps still not quite yet ripe. One can have, in the present situation, perhaps more flexibility to see how competition develops and what is needed as a result of that. But as time goes by the desirability of more transparent and better formalized arrangements will become clear and I certainly believe that that position will have been reached before the year 2000. I emphasize that it is not a break-up of BT, but a separate accounting division that I envisage.

On the subject of technology, I speak with total diffidence in the sense that I am not a technologist myself and while I enjoy learning about it from others, I can claim no expertise in it at all. Indeed, I am not even interested in predicting what technology will prevail, because I want to stress that a regulator should *not* have the objective of picking a particular technology as a winner, and then trying to enable it to become one.

Instead a regulator should aim to create a situation in which *all* possible technologies have a reasonable chance of developing, so that there is no danger of cutting off progress by suppressing the technology that turns out ultimately to be the most promising one. Of course, I recognize that in the industry of telecommunications chance matters of timing can have an undesirably great effect on which technology triumphs. If you have undertaken a big 'lumpy' investment, like putting up a satellite just before some other technology for providing television comes established, then that creates a different situation from the one in which the other service came first because you can't bring a satellite down again and recover the investment you made in it, or even a large part of it.

Nevertheless, and accepting that imperfection of the market situation, it does seem to me to be a very important aim. Indeed a regulator should have the main technologies in sight and to try to give each one a reasonable opportunity to develop. If you look at the growing situation in the UK you see a very interesting set of possibilities and some permutation of these perhaps will shape what we have in the year 2000. We see direct broadcasting by satellite for television vying with the now growing cable television industry and traditional off-air broadcasting. Which one of those will be the service that dominates in the future? You see in the basic telecommunications field, a contest between cable and radio. My projections at Oftel do suggest that mobile networks, as they expand over the next ten years, will be able to provide service to ordinary residential customers at about the same level of cost as fixed networks. It will become sensible economically, if it's sensible for other reasons, for people to decide that they will use their mobile phone at home and when they are out as well, that will be a cost-effective solution and will, of course, provide a high level of service.

And in addition to that, we have the possibility of providing service by radio over fixed links, rather than mobile links. Now how will these things work out? It will be possible for people to have satellite services and mobile radio for telephone and not have a cable going into their home at all. I would not want to predict that situation will happen, and although I do not want to predict anything in this particular field, I will share with you my suspicion that it will not actually happen. I do think that mobile radio service will become universal. As I like to say: everybody will have a telephone in their pocket or in their handbag before too long. Perhaps we will be about half-way there by the turn of the century, but I suspect mobile service will increase rapidly again over the next ten years. I suspect also that broadband cable into the home will be the way of the future. I think the cost of having both will be sufficiently low eventually that people will actually opt to have both and enjoy the greater interactivity that you can get through cable, compared to satellite.

We may very well find that we eventually get the position where cable and radio have traded places: where radio is not used for broadcasting television pictures any more, but for providing telephone service, and cable is used for providing a television service and the radio waves are not. I emphasize again in expressing the objective the way that I do, that the aim is to create opportunities for the deployment of technology. But I do not say that, believing the market forces that will then take over to be perfect. I just believe that the governments and regulators are even more imperfect in picking the technologies that will prevail than markets are.

Regarding traditional approaches to regulation, the last ten years have seen a pioneering in the UK of incentive approaches to regulation, in which the regulatory devices used encourage greater efficiency and better customer service through their properties. A price cap regulation is a good example of that. You set a ceiling on prices but you don't control profits *per se* and you allow the regulatee to make more profit, if it can do so, by becoming more efficient than you expected in setting the price cap in the first place.

I believe that price cap regulation has been successful and I believe that it will continue to be used, not only in the UK but by more and more countries elsewhere. I see interesting signs that some other European countries are thinking of using this approach now, as well as of course the US who have adopted the approach to some extent. To me, the logic of the approach is so obviously right that I cannot believe people will move away from it. Of course you can reduce profits to a low level, if you really want to. But the cost of doing that is likely to be higher prices for service since the regulatee no longer has an incentive to become efficient. Without such incentives, people probably won't become efficient, especially if competition cannot be made to bite hard, and quickly as well.

I see this kind of arrangement as having been successful also in quality of service. The Citizens' Charter idea about the provision of good service has been the subject of another pioneering effort in telecommunications. BT's pioneering Customer Compensation Plan, with of course other factors, has produced dramatic improvements in quality. Overall, although a regulator has to insist that he is looking for 'more and better', and I do, I have been very well pleased with improvements in quality of service over the last few years, and I think it is right to emphasize that. I expect those improvements to continue. I expect the fault rate, in particular, to decrease sharply by the end of the century; It seems that the performance in terms of fault repair and provision of new service, is a great deal better than it used to be and is well on the way to achieving a satisfactory level.

My final prediction is that Oftel, or some other regulatory body at least, will still exist in the year 2000. All these good developments in the form of competition will not have gone so far by the year 2000 that regulation will no longer be needed. Some people have suggested that my objectives should be to work myself out of a job by promoting competition so strongly that we do not need regulation any more. I yield to none in my promotion of competition and yet, I simply do not think it is realistic to imagine that competition will have reached the point by the year 2000 that regulation can be abandoned. Two competitors in a particular section of the market place is not enough for that; you need at least three, and you also need it on a widespread basis. So, my final thought is that my successor will still have a job to do in the year 2000.

9

Communications strategy – the impact of market forces, regulation and technology

N.W. Horne, Partner, KPMG Management Consulting

9.1 INTRODUCTION

Every day throughout the world millions of people travel into cities to work. Every day throughout the world millions of people move their homes closer to major conurbations so that they can join the existing millions in travelling into cities to work. As manufacturing becomes less labour intensive the pressure to find work in service industries based on cities increases.

The result to the present way of doing things is to produce congested, polluted cities, hours of costly, uncomfortable commuting, high overhead costs for business in providing expensive city premises, massive consumption of scarce natural resources for transport, worldwide environmental damage as a result of the energy consumption, disruption to the lives of more millions to build more road and rail links to enable more commuting, and depopulation and skill starvation outside cities.

The problem is bad enough within national boundaries. With the proposed economic integration in Europe the danger of concentration of population (with all the consequences given above) into the so-called 'banana belt' is acute. There has to be a better way.

Communications has a key role to play in this situation in exactly the same way as it has had in the development of civilization in the past. The massive capacity available in fixed, fibre optic networks coupled to the versatility of mobile communications will change the way in which 'white collar' tasks are performed as well as provide national and international information and services wherever required. Many of the prob-

Communications After AD2000. Edited by D.E.N. Davies, C. Hilsum and A.W. Rudge. Published in 1993 by Chapman & Hall, London, for The Royal Society. ISBN 0 412 49550 3

lems of present day cities can be tackled and there are real economic incentives as well as the need to address the issues of environment, quality of life, conservation of natural resources and other factors.

The idea of a high capability communications is certainly not new. The idea of the potential benefits to be obtained was first realized eight or ten years ago. The idea is no worse for that. What has been missing over the last few years is the appreciation that technology can be used to improve and quicken the processes of deregulation and privatization and improve services to the customer as well.

The technology of communications can play its part only if the regulation of the market allows. A new strategy to regulation is required to take account of what technology can do. The emphasis has to shift from attempting to provide existing well understood services more cheaply, to encouraging investment in networks which will enable real competition in service provision as well as provide the infrastructure for a new way of life and work. The starting point is to separate services from the networks needed to provide them.

This paper describes the market forces which will cause the installation of high capacity fixed networks to every home and business location in the third millennium coupled to improved radio networks for those on the move. It will also describe some aspects of the technology and the changes to market regulation which are required to make it happen.

9.2 MARKET FORCES

Conventionally, market forces are measured in terms of demand, the effects of competition and so on in the short term. It is assumed that the supply side of the economy will react to satisfy the market forces. In the case of telecommunications this model breaks down for two reasons. The first is that few market demands can justify the infrastructure investment required to supply them. The second is that the market is not 'free' – it has to be subject to detailed regulation because of history and because there cannot be too many suppliers of services if necessary investment is to be encouraged. In this case some of the market forces are of the demand type but some are of a more long-term and global nature – such as environment and political issues. Seven issues will be considered: political forces in Europe, National costs, company profits, environment, energy conservation, quality of life, and technology.

9.2.1 Political forces in Europe

The integration of Europe will place demands on all forms of administration to co-ordinate and integrate their functions without building up 'federal' functions. Obvious examples lie in law and order, where, as the barriers to movement are removed, it becomes essential to interwork police, customs and so on as never before.

Less obvious examples occur in the fields of media communications, reduction of language barriers, equality in education, equality and consistency in health services, the provision of pan-European information services. Perhaps equally important is the need to avoid depopulation of the deprived areas of Europe as people gravitate to cities to find work.

Communications is an essential part in addressing these forces as it has been in the development of civilization. The faster, better and more far reaching the communications, the more effective it will be in making a contribution.

9.2.2 National costs

Every government worldwide is faced with ever-increasing costs in administration and building infrastructure. Most governments have a large proportion of their employees in and around major cities – there are about 350 000 civil servants in public administration and defence who work in London. Major city costs are high and there is a constant need for renewal of premises – the Department of the Environment recently announced the replacement of their premises in Pimlico, only about thirty years old, and their replacement at a cost of over £200 million. These costs translate into more taxes.

Apart from civil servants, governments have to support the commuting of millions of others – at least one million people commute into London every day – and provide the necessary transport facilities in roads, railways and motor transport. This is not only costly in itself – there are 16 000 civil servants in the UK planning the road and rail system – but costs are passed on to industry of all kinds in congestion – the CBI have described London congestion as 'the business issue of the 1990s'. It also incurs more costs as people are forcibly removed from their homes to make way for new works.

Every nation has to incur the energy costs for commuting and congestion and suffer the resultant deprived and depopulated areas which result from the migration of the population to cities.

It is conceivable that better communications will remove the need for some of this commuting. Investment in new and better communications infrastructure might give a better return than more investment in road and rail infrastructure.

9.2.3 Company profits

Business and individuals have to bear the cost of all this in the form of higher taxes. But there are more direct costs. Accommodation costs in

London, including all the necessary services, can be £150 per square foot per year or even more. If we take an average sized services company of say ten employees and assume some average statistics (say £100K per annum turnover per employee, 70 square foot of space per employee, 10% profits on sales) then profits will be about £100K per annum – and accommodation costs will be about the same sum. This means that profits could be substantially increased and/or the cost of services substantially decreased by not being in central London.

As if that is not enough, the cost of congestion is high. The average speed of traffic in London is 11 miles (16 km) per hour. The CBI estimate the cost of congestion throughout the UK as being £15 billion per annum (£10 billion in London). British Telecom estimate that a 1.4 miles per hour increase in speed could save them £7 million per year! The situation is getting worse. In December 1987 there was a 7 hour standstill in part of central London as 'gridlock' caused by road works and heavy traffic brought everything to a halt. If something is not done the consequences are becoming clear.

9.2.4 Environment

Thousands of cars, motorcycles, buses and other forms of transport are used every day to transport people to and from work and as part of their work – in London about 1 million commute in total of which about 180 thousand use private transport. In central London motor vehicles account for 1.05 million tons per year of pollutants (out of 1.3 million tons) and are the dominant source of carbon monoxide, nitrogen oxide, hydrocarbons, smoke and lead. It is estimated that every car in the UK – there are over 2 million in London – produces four times its own weight in carbon dioxide alone each year. European Commission limits for nitrogen dioxide from traffic fumes were broken in the late 1980s in the London Boroughs of Hillingdon, Ealing, Hounslow, Brent, Kingston, Newham, Southwark, Hammersmith, Fulham, Westminster, Camden, Islington, Tower Hamlets, Lewisham and Greenwich.

As if this is not enough, major cities have concentrations of heating plants, air conditioning plants, emergency power diesels, waste disposal plants and so on which add to the problem. Temperature inversion is seemingly becoming a more frequent phenomenon, trapping pollution over cities. Pollution in many cities is reaching such a level that the only way to reduce it is forcibly to prevent people from living and working there.

It would be silly to suggest that all travel to do with work can be eliminated, but not so silly to suggest that a broadband communications network, able to carry every conceivable form of communications from multiple voice, colour facsimile, and computer files through to high definition colour pictures, multiple high definition colour television channels, can eliminate some, and perhaps the majority, of work travel.

9.2.5 Energy conservation

Working from the 1988/9 National Travel Survey from the Department of Transport, there are about 170 billion passenger-miles of car travel in the course of work undertaken in the United Kingdom each year. If one takes average fuel consumption at about 25 mpg (remembering the very slow speeds achievable in towns and cities) then this amounts to up to 7 billion gallons of fuel consumed. This fuel is not only expensive, it is irreplaceable. Of course, no communications network, however good, can replace all of physical travel but it can replace some of it. We surely have had enough fuel crises in recent years to know the value of conservation of energy for essential support of our civilization.

Effective communications, by reducing the need for travel, can contribute to energy conservation.

9.2.6 Quality of life

We have already mentioned the congestion and pollution now being experienced in our cities. The effect on the quality of life of our citizens is severe. If one considers commuting alone, most people spend an hour of more travelling each way in crowded, dirty, unreliable public transport or trapped in congestion using private transport. Apart from the frustration and stress, people are subjected to considerable indignity and cost – season ticket prices between £1000 and £2000 are now commonplace. Anything that will reduce commuting and yet allow people to make the same contribution to wealth creation and life in general has to be welcomed.

But quality of life is not just about the removal of negatives to do with work. It is also about avoiding all unnecessary travel and living where it's nice to live without being isolated and cut off. It is about recognizing the importance of uniform education, not just in schools, but education which will allow people to have two or more careers, learning new skills when it suits them. It is about access to information from wherever one is, access to news, libraries, statistics, general knowledge, business trends, market reports – even consumer data. It is also about entertainment on demand when required not just when programmed by someone else.

A high capability communications network can contribute to all of these but would probably not be cost effective for any one of them.

9.2.7 Technology

Technology is, itself, a market force. The relentless development of capability in communications, processing power, memory, distributed computing and network management continues apace. The pace of technology forces the use of a common network by many different services. After all, in terms of digital technology, whether a bit stream represents voice, data, image or video is totally transparent in a fibre optic cable.

Communications has always been technology driven and has always been a long-term business where technology change has to be planned years in advance. Technology allows the regulator and others to exploit it to achieve such aims as competitive supply of services and networks as well as having flexibility in how many services can be supplied in the future.

The latest developments in fibre optic technology potentially allow an undefined number of services of all kinds to be provided, using wavelength division and conventional multiplexing, with comparatively cheap combination and separation of services. Passive optical networks can be protocol transparent and light can be amplified using optical amplifiers which are also protocol transparent. All of this gives great flexibility. Of course, for any one installation the usual decisions have to be made and some of the flexibility will be removed in that instance. Nevertheless, the basic flexibility can be exploited and will be exploited as the technology and capability advance.

Regulation must always anticipate technology advance if it is not to confine users of communications to outdated services. The emphasis in regulation which has concentrated on obtaining competitive supply of voice telephony (with only mixed results in the fixed network) must now shift and embrace and steer technology and investment to achieve real competition in voice and advanced services for the future.

9.3 HIGH CAPABILITY COMMUNICATIONS

9.3.1 Communications infrastructure

All the problems and opportunities above lead to the conclusion that a high capability communications network can not only make a contribution but is an essential feature of life in the third millennium. What is needed is a very high capacity network connected to every home and every desk in business premises, coupled with high capacity mobile radio for people on the move. These would act as a 'common carrier' to services of all kinds. When high capability multi-function terminals, capable of dealing with voice, text, image and video are attached then, in principle, all of the market forces described above can be addressed at least to some extent. This is not just an extension of the present telephone or cable television networks. It is a quantum increase in capacity so that, for example, hundreds of high definition video channels can be made available to every person to be used for work conferencing, selling, entertainment, shopping as well as voice and data channels for telephony, computer text files, mail handling, information accessing.

The capability of such a network and its consequences need to be carefully thought through. If the architecture is right, it will be capable of providing multiple voice channels giving true competition in public telephony for the first time – even after eight or nine years of 'deregulation' in the UK most people have not been directly affected by competition at all. It will be capable of carrying the services of many cable television providers thus giving real competition – unlike at present where only one provider is licensed in each geographical area. It will be capable of carrying every service required for business, from voice, letter delivery, text access to computer files through to multiple high definition television conferencing, remote operation of equipment, access to business information. It will be capable of monitoring and remotely adjusting activities in the home. All at the same time and in a fully co-operative way, not just by sending messages but by jointly and simultaneously working together.

In short it will provide 'distance independence' – the ability of work and play worldwide regardless of location and geographical distance – for the first time. It will improve all aspects of our lives but will not, of course, be any substitute for the normal social intercourse on which every human being depends. The difference is that meeting and socializing, as well as work, can be done where it is nice to be when it is nice to be there and not in cities and other congested areas.

9.3.2 Addressing the market forces

Such a network can make a major contribution to the needs of the market. Political integration in the working of administrations in Europe can obviously be improved by the ability to interwork in voice, text, image and video in any combination. The provision of education, information services, health documentation anywhere and at any time will give more equal opportunities to Europe's population. The possibility of avoiding depopulation by allowing work to be brought to the person rather than the person to the work is attractive. Much of the cost of the network is being expended already through multiple networks in the local loop of inferior capability (telephony, cable TV) and any increased costs could well be offset by avoiding costs on expanding public transport and on roads.

Most government employees are white collar workers where, at least in principle, the work can be taken to the worker. The possibility of establishing smaller work centres, widely distributed throughout the country, if not actually working from home, could lower costs, reduce the need for large office premises, reduce commuting and lower pollution.

Company profits can be improved and/or the cost of services reduced by keeping white collar workers out of big cities and avoiding the resultant costs of congestion and uncertainty in providing customer service.

The environment can be improved by such a network. The reduction in unnecessary travel, the avoidance of concentrations of cars, air conditioning plants and so on will reduce city pollution. Similarly, reduction in unnecessary travel will reduce energy consumption.

Above all, such a network will enhance the quality of life for us and our children. Avoiding the waste of time, the indignity, uncertainty and costs of commuting unless absolutely necessary will be a major contribution. Living and working where it's nice to be and breathing cleaner air will add to that. Having the opportunity of uniform education throughout life, access to health records, libraries, statistics, shopping, entertainment when needed will take it still further.

Finally technology will *force* the introduction of such networks as it relentlessly develops. Each nation must decide when to introduce the capability. Being too early will cost a lot and may quickly be outpaced by technology. Being too late will deflect investment into unsuitable networks, decrease competitiveness, provide incentives for good people to go elsewhere. But a network has to be of the right architecture to fulfil the regulatory and market aims. Just because a network uses fibre does not mean it is satisfactory – even if it connects every house and desk in the land. What is required is a positive lead to technologists to encourage them to develop networks with the characteristics most desirable from a service capability, flexibility, future proofing and competitive point of view. Out of all the vast portfolio of capability available this would focus development. At the very least every nation must avoid encouraging investment which will constrain the introduction of such networks by building vested interests in more limited approaches such as cable television, the costs of which have to be recovered.

The implementation of a high capability network does not mean the need for large amounts of government money. The implementation of the network can be achieved competitively or centrally, provided the appropriate standards and interfaces are properly defined.

9.4 DEFINING COMMUNICATIONS STRATEGY

The vision of a high capability network, coupled to high capability work stations seems highly desirable, particularly as we shall show not only that the basic technology is already available but the costs of such a network are already being borne, at least in part, at present in the UK in spending on old fashioned networks of much lesser capability. What then is a communications strategy which might fulfil the vision?

9.4.1 Services

First it is about the competitive procurement of services. The end-user of communications networks does not know or care about the network itself. What he or she is interested in is what can be done with the network – from voice telephony to multiple video conferencing. Since no-one can forecast the variety of services which may be needed and supplied, the network must not depend on the type of service. It must be service independent to the greatest extent possible. The provision of services must be separated from the provision of the network and there must be well researched and clearly defined, simple interfaces between each service and the network.

Each user of a service must be capable of being accessed by an address, not just as an individual but also to indicate which service is to be used. At present the telephone number acts as this address for telephone service. We need an addressing scheme which allows an effectively unlimited number of services, including multiple telephony providers, and which will find an individual wherever he or she may be. Since we are dealing with distance independence the address must not depend on the physical location of an individual. The method of addressing in our strategy needs to be re-thought from scratch. The present method of generating different totally unrelated numbers for telephone service using British Telecom (BT), telephone service from Mercury, facsimile service from BT, facsimile service from Mercury, cellular telephone service from BT, cellular service from Vodaphone, PCN service from a number of providers, telepoint service, cable television service and so on simply is not adequate. Once the basic idea of many services using the same network, and the idea of distance independence, are grasped then it is clear that a structured numbering scheme is required. This can be designed to allow new unknown services as well as providing a logical pattern. A possibility would be to preface or follow a personal number with a code to represent the service to be used.

Above all, the strategy must encourage long-term investment. Communications is a long-term business – for users and providers. The present approach of devising schemes which require users to change their numbers every few years, schemes which encourage the installation of networks optimized for a single service, such as cable television, schemes which require investment in new networks before new services can be provided, must be avoided.

9.4.2 Networks

The second part of the strategy concerns the provision of networks. We have seen that this provision must not be tied to the provision of services. The investment required in high capability networks is large. It is estimated that the cost of providing a local network – from the local hub to the house – of any kind to any one house can be £350 or more and most of this is labour cost in digging trenches and the like. At the present time in the UK we are spending such sums on the provision of cable television networks and on renewal of British Telecom local cable network. BT has been prevented from carrying other services such as television on its local network by regulation so that its spending has been at an all time high on copper twisted pair in recent years as BT improves its service. What this amounts to is investment in history.

What is needed is a strategy which, instead of encouraging these investments in old technology, encourages investment in new technology which really provides competition, and which can be done for a similar cost. The strategy must not be dependent on how new investment in networks is funded. It must accommodate private, competitive funding as well as central government funding. It must also be independent of whether there is one such network carrying all services or many networks with the services shared between them.

The best approach is to assume competitive procurement of networks. The requirement then is to research and define interconnection standards such that all network providers can interwork. These standards are not just simple 'plug and socket' but must recognize the end-to-end network management required, the national coverage required and the need to interconnect through gateways to international capability. The requirement is for every piece of network to be able to carry any or all of the services and to interwork with every other piece of network.

A key feature of such standards is that they must integrate the needs and capabilities of communications and computing. Standards which are set, say, for voice without considering data needs are not adequate. This is a matter of administrative and regulatory organization. The separation of telecommunications and computing must be removed both nationally and internationally if satisfactory standards are to be developed.

These are not easy issues and a few years ago would have been seen as impossible. Today, the enormous strides in distributed computing, distributed databases, distributed network management, all make such end-to-end management of the networks and the services a possibility. The need now is for focused research and development to overcome the perceived difficulties.

9.5 STRATEGY IMPLEMENTATION

The implementation of the communications strategy requires three main activities – derivation of the appropriate rules and standards (regulation), derivation of the appropriate technical solutions (technology) and organizing to ensure and respond to the opportunities (administration). The first of these is regulation.

9.5.1 Regulation

The prime need is for a statement of goals or vision as described in this paper. The central core of such a vision is real competition in service provision and network supply anywhere, any time and at reasonable cost. Clearly, if networks and services are to be allowed to be supplied competitively then, by definition, any network must be allowed (indeed enforced) to carry any service – there must be no restrictions as at present. This then leads naturally to the requirement to specify the interfaces between services and networks and between the networks themselves. Much of this will be amenable to international effort, particularly in Europe. Regulation will then be concerned with enforcing standards and ensuring fair charging for services to end-customers and for common carrier network transport to service providers.

Finally, regulation is about defining and enforcing a suitable addressing standard which will anticipates future needs for many years to come. No single network nor service provider can derive such a scheme – it has to be derived by an independent body and, again, may best be settled internationally.

9.5.2 Technology

Most of the basic technology required for the high capability communications network is already available or capable of development. The fundamental structure would be a high-capacity fibre trunk network coupled to fibre to the house and desk. The technology in the trunk network can be synchronous digital hierarchy (SDH) or any other management scheme with high capacity and flexible reconfiguration. In the local loop the most likely technology is passive wavelength division multiplexing since that is the simplest way of providing many independent channels with cheap and easy combination and separation at each end. The theoretical capacity this network could provide could be as high as 1000 Gbits s^{-1} to the desk – most of which would be unused by design – compared to the present ISDN standard of 144 Kbits s^{-1}. This fixed network would be coupled to digital high-speed mobile networks capable of taking any service easily.

The network terminals would be high-capability workstations coupled to telephones such as are available today. Their processing capability would probably be up to 1000 MIPs (million instructions per second) with plenty of memory – disk memory now is as low as $10 per megabyte on large disks and continues to fall. The terminals would eventually have large-area colour flat-panel displays.

The key to success in terminals is not so much in the hardware but in the software. The current developments allowing distributed management, computing and databases are key.

Looking at the toolkit of technology presently available and the work being undertaken to improve it, the most striking thought is that if there was a clearly defined objective to direct development then the capability required for a high capability communications network of the sort described in this paper could be much closer than most people think. What is needed is regulatory and market *pull* to make the technology solve the political, cost, profits, environment and quality of life issues covered earlier.

9.5.2 Administration

The third part of the strategy concerns administration. It is clear that for services of any type – voice, text, image, video, or any combination – to be provided on a common network then the standards and interconnections required are not solely concerned with telecommunications or the network itself. The standards must include interworking, distributed management, distributed databases and so on – standards which traditionally would have been considered as computer or information technology standards.

This means that to derive the standards there must be a single unified approach between telecommunications and computers. The separation which occurs in many government departments (including the European Commission), companies, universities and other bodies has to be broken down. The eventual need for such a move cannot be in doubt but the timing of it is difficult. Government must work towards such integration of functions and regulatory responsibilities quickly to achieve the benefits of a modern network infrastructure.

9.6 WHAT NEEDS TO BE DONE

- First, we need to plan future communications strategy in recognition of the potential savings which can be made in other forms of infrastructure such as roads and rail transport.
- Second, we need to plan future communications strategy taking

account of the contribution which it can make to European integration, national costs, company profits, environment, energy conservation and quality of life.

These will provide challenges to the working together of government departments. We need to do both of them in recognition that the prize is not only to address the market forces but also to gain true competition in communications services.

- Third, we must specify the requirements for competition, flexibility and future proofing required and encourage the development of network architectures which meet them.
- Fourth, we must invest in the derivation of standards and interfaces to allow networks supplied from various operators to interwork and to allow the transport of any service on any network.
- Fifth, we must derive a network-independent addressing scheme which will allow for a wide variety of services in a logical fashion.

Most of the investment required for these can be substitution for existing support for research and development in the UK and in Europe. The focus which will result for such programmes as the Joint Framework for Information Technology in the UK, ESPRIT and RACE in Europe will be a great benefit in pulling technology through to good use.

- Sixth, we need to licence network operators and service providers in the knowledge that future licensing will separate the two so as to discourage investment in old fashioned network architectures.
- Seventh, we should put a requirement on network operators to carry all services throughout their networks as soon as technology allows.
- Eighth, we must pull together the administration of research and development and of standards in telecommunications and computing in the UK and in Europe.
- Lastly we must recognize that developments in technology will relentlessly continue. If we recognize them and take advantage of them we will ensure the position of the UK in the forefront of communications infrastructure. If we ignore them we will be disadvantaged for decades to come relative to other countries.

9.7 CONCLUSIONS

Such are the developments in technology and in market forces that a communications strategy based on trying to force-fit existing networks into competitive environments, defining networks optimized for specific services, modifying numbering schemes with minor adjustments, will not suffice for the third millennium. Tomorrow's demands will not be satisfied by doing what we did yesterday with minor changes. We need to think through a new approach which will be flexible in terms of competition, new services and change in order to encourage investment and confidence. We need to invest in the development of standards and interfaces to allow for the flexible provision of services and networks.

There is almost too much technology available. What is needed is direction to development to produce network architectures which allow almost unlimited competition in services, that are flexible, adaptable and consistent in application and addressing.

Market forces today demand a new approach. If we are to provide a safe, economical, satisfying, competitive life for our children, we must have a new approach. This paper outlines how a communications strategy for the third millennium is both essential and achievable.

10

Architecture, technology and applications

M. Ward, Technical Director, GPT Limited

Can we predict the technologies that will provide the foundation for telecommunications networks and services in the first ten years of the next century? Well, . . . it depends. Are we concerned with the spatial and geographical aspects of telecommunications: with 'bridging distance'? If so, the prediction is relatively easy. Look for the technologies that will emerge from the laboratories into pilot service trials, say, in 1993.

However, if we are more concerned with the use of technology to ameliorate the deficiencies of the distance network or to enhance its basic spatial properties – to 'adding value' – then its deployment is much less predictable. We must look for those technologies which will be emerging from the laboratories in, say, 1998.

This paper will analyse the 'bridging distance' and 'adding value' domains of telecommunications in more detail, in order to predict the likely shape of telecommunications networks and services in the next century.

10.1 INTRODUCTION

It can be claimed, with some justification, that the largest and most complex machine in the world is the international telephone network. Without doubt it is large, reaching to most parts of the world's surface and a fair proportion of its inhabitants; it is a machine, in that its component parts work together automatically to achieve the desired results, and it is complex. This latter point can be illustrated by considering, for example, what has to work together and what technology is involved in sending a fax from someone in the UK to someone in the US via a satellite link.

Communications After AD2000. Edited by D.E.N. Davies, C. Hilsum and A.W. Rudge. Published in 1993 by Chapman & Hall, London, for The Royal Society. ISBN 0 412 49550 3

The telecommunications community can take some pride in having been involved in the design and construction of such a vast machine. A mark of its achievement is that in many parts of the world the machine is taken for granted, while in others the penetration of the telephone service is used to calibrate the economic development of a particular community. Yet its very ubiquity together with its complexity can act as a barrier to the deployment of new technology and the emergence of new services.

This paper will attempt to provide a framework for a discussion of the way network architectures influence the introduction of new technology. It will then examine three applications of technology in the network and conclude with a survey of some of the services which could be deployed as a consequence. This discussion shows that there are two forces at work: one which seeks to extend the functionality of the network and the other which seeks to exploit, to make the most of what is already there.

10.2 BRIDGING DISTANCE AND ADDING VALUE

In order to establish a framework for the analysis, we can divide the 'science' of telecommunications into two domains. The first domain encompasses the spatial and geographical aspects of telecommunications: bridging distance. The second concentrates on the use of technology to overcome the deficiencies of the distance network and to enhance its basic spatial properties: adding value. Here, I refer to the first as the distance domain and the second as the value domain.

The distance domain includes the network required for bridging distance between communicating entities and the functionality required for setting up the communication channels (connections) and for operating this network efficiently. Investments in this distance domain are massive, long term, concerned with infrastructure and are often highly interactive. One result of these factors is that there are comparatively few players in this domain. Another result is that the introduction of new technology takes a very long time. Technologies that will form the basis of the distance domain in the first decade of the next century will be emerging from the laboratory now.

The value domain on the other hand is concerned with the short-term enhancement and exploitation of the characteristics of the distance network. This domain often deals with needs which are very specific and which may vary over a short period of time. Our example of this is the enhancement of switching equipment by intelligent network (IN) technology to create the freephone and premium services. Another is the exploitation of the 3 KHz channel by facsimile terminals.

These two domains can be described as being in competition with each other. For example, the distance domain advertizes the benefits of

ISDN – end-to-end dial-up 64 Kbits connections – in order to justify the expenditure entailed in its introduction. While this whets the appetite of the user, the deployment of ISDN is an infrastructure development and takes a long time. So the potential users attempt to satisfy themselves by further exploitation of the ubiquitous 3 KHz channel. Hence high-speed modems on the one hand and signal compression on the other.

While the obvious way of developing services on the broadband network would be to extend an optical fibre to each user, this would be too expensive and take too long. In addition, demand for services often arises at random points over a highly dispersed population, while the deployment of optical fibre networks has to be planned. So some engineers are investigating how much more capacity could be squeezed out of the ordinary copper pair and others are devising yet more algorithms to further reduce the transmission requirements of high quality video communications. A third group would use satellites for local access.

10.3 THE ARCHITECTURE OF THE DISTANCE DOMAIN

There is a general familiarity with the architecture of telecommunications networks in terms of nodes and links. From the architectural point of view the function of the links is simple: to get information from A to B as cheaply as possible with no degradation of quality and maximum reliability. I would want to maintain that because the function is simple, the exploitation of the latest technology is, relatively, easy. Optical technology has matured in less time than it has taken to connect 50% of UK customers to digital switches. Indeed, the very latest optical technology is being used in what seems at first sight, to be the most unlikely situation: the Atlantic and Pacific cables.

Perhaps the essential principle at work here is one of transparency: as long as the same information is carried, the engineer is free to exploit the latest technology to achieve economies of scale and to improve reliability and quality. If this same new technology can also be used to increase the scope of the link, for example, to add transparent 64 Kbits ISDN to the 3 KHz analogue channels, then so much the better.

The principle of transparency can also apply in a different way. If, once the basic link is established, its capacity can be increased merely by upgrading the terminal equipment, then the period between conception and exploitation of technology can be further reduced. This can be illustrated by single mode fibre where it has been possible to upgrade links from 34 Mbits to 140 Mbits and now to 565 Mbits, merely by changing the terminal equipment. The enemy of this form of enhancement is the digital regenerator, particularly when it is buried thousands of metres under the ocean.

While the functions of the links are relatively self-contained those of the nodes, the switches, are highly interactive. Nodes have to co-operate with each other in order to route and switch calls to their ultimate destinations and this interaction can be complex. Modern switches are connected together, not just by channels which carry the customers' information, but by high-speed data links which carry the signalling messages required to set up and clear calls. These signalling messages allow the software in one switch to talk to the software in another and it can be seen that if the content of these messages, the meaning of these signals, is changed then the whole machine may fall apart. Thus if we are going to introduce a new switch into a network it will have to perform all the functions that are performed by the existing model. Again, if we wish to enhance any existing switch, say to increase functionality, we have to ensure that, as far as the network is concerned, it continues to operate in the same way as before in addition to performing the new functions. Similar arguments apply to the 'change out' situation where a switch is replaced by an new version with enhanced functionality. In this example the old and the new may have to work in parallel in order to maintain service to the customer during the change-over process.

The worst problems arise when changes are required to the way the switches talk to each other; when the meaning of the messages between the software in each switch changes, when the semantics of the network shift. For example, parallel versions of the software may be required to accommodate the old and the new in the change-over period. Software has to be tested even more rigorously than usual so that the whole network does not collapse when the final cut-over occurs. And if there is more than one type of switch in the network the problems can increase by orders of magnitude.

The functional complexity outlined in the preceding paragraphs indicates why the nodes of the network can lag behind the links in terms of the introduction and exploitation of new technology. However there are certain architectural techniques which can be used to reduce the difficulties and speed up the process of which the most powerful is overlay.

The most obvious example of this is the satellite: in their most radical form satellites can be used to by-pass the complete network in order to introduce a service which cannot be provided by what is already there. They can also be used as partial overlays, for example, where it would be impossible to justify, in economic terms, the upgrade of the copper local loop but where broadband access to a particular location is required for a special new service. Nevertheless, the customer will have to be prepared to pay.

Consider the introduction of a new service. The transmission technology already installed is quite capable of providing the right bandwidth, but new switches are required. In this case the existing transmission capability can be used to by-pass the existing switches in order to access a new switch provided as an overlay at some point in the network. The difficulties arise of course when some interaction is required between the old and the new. Overlays can also be used to enhance the quality of what is already there. During the process of replacing analogue by digital switching in the UK, the new digital switches were introduced as an overlay, initially to absorb traffic growth and finally to replace the old equipment.

To conclude this part of the discussion, it seems to me that answers to the following questions could help us to predict the introduction of new technology into the distance domain of telecommunications.

- Is the technology there and can it be brought to an appropriate stage of completion in a reasonable time?
- Are there potentially attractive new services which will need this technology in the sense that the service cannot be provided by other means? A supplementary question here is whether the service could be implemented by the application of different technologies in the value domain. For example, let us consider on demand video. While this is one of the services which is used to justify the deployment of fibre in the local network it might also be possible to implement an acceptable service on the existing network by compressing the signals to remove redundancy and by utilizing new developments in signal processing to get more out of the copper local loop.
- Does the technology affect the node or the link (or both)?
- Can the technology be applied as an overlay?
- Will the existing structure be sufficiently transparent to apply new technology and will the new technology be transparent to existing services?
- Can the principle of super setting be applied: can the new technology absorb what is already there?

We will now examine some technology trends in this context.

10.4 ATM

ATM stands for asynchronous transfer mode and first emerged into the light of day in about 1984; it has been the subject of world-wide activity ever since. While only laboratory models have been demonstrated so far, the pilot network trials are planned for 1993 and 1994. The long timescales provide a good illustration of the difficulties of introducing radical new technologies into the distance domain.

The basis for ATM, as indicated in Figure 10.1, is a high-speed digital

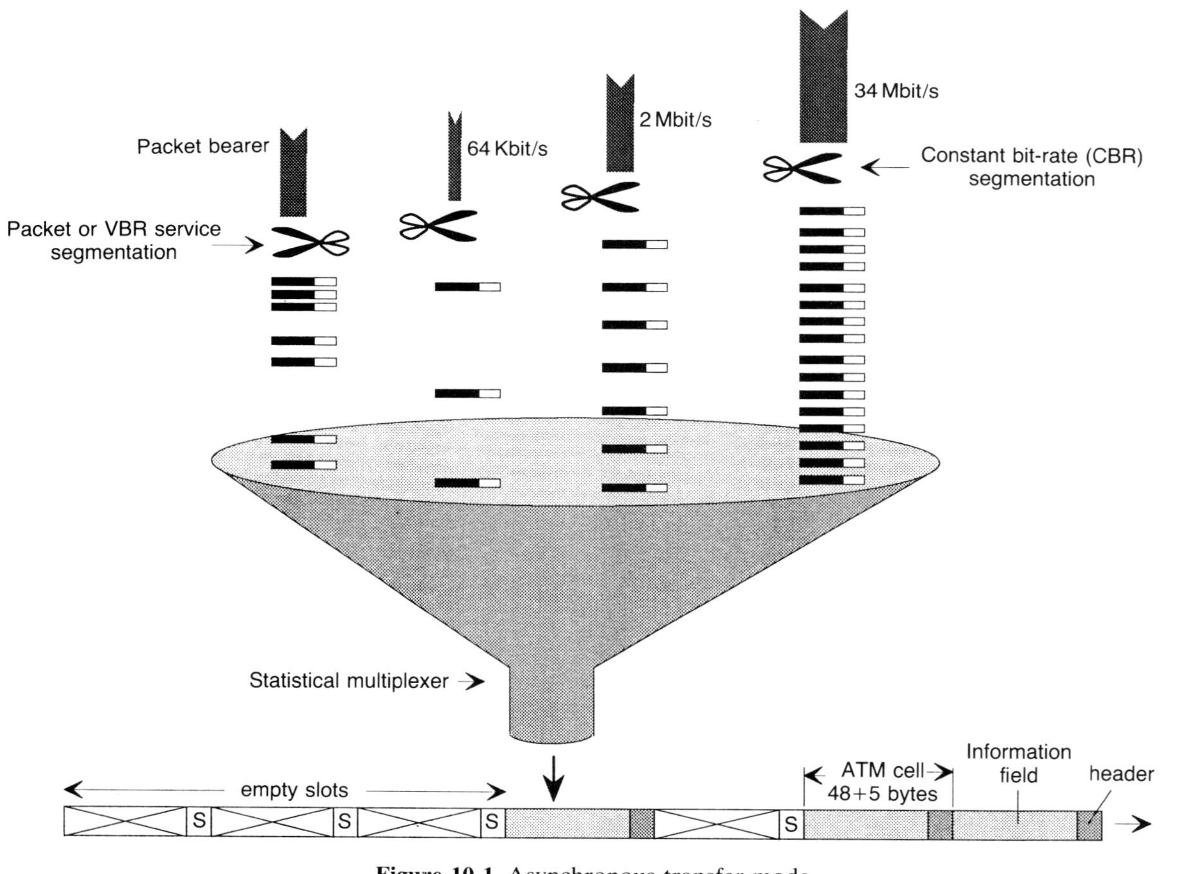

Figure 10.1 Asynchronous transfer mode.

bearer which is divided into fixed length cells. Each cell has a 5-byte header, which contains information about the 'call' to which it belongs, and 48 bytes of data. For the purposes of this description let us assume that, initially, all the cells are empty.

The funnel in the diagram represents the process of filling cells with data and is somewhat similar to the familiar process of quantization: the aggregate cell stream, at the bottom, has to be fast enough to carry the various tributary streams transparently, without distortion. Each tributary stream is subjected to a process of division into fixed length cells, represented by the scissors in the diagram. When a cell is full, it is given a header which, *inter alia*, provides information about its destination, and it is then dropped into the funnel. This acts as a large first-in-first-out queue, where the point of entry into the queue is governed by when the scissors close; that is, when the cells are full.

Although the ATM funnel bears a resemblance to statistical multiplexers in a packet data network its purpose is fundamentally different: it is to multiplex input channels which arrive in a variety of different forms into a single, consistent medium. The input channels can be 'constant bit rate' (a 2 Mbits primary mux), 'variable bit rate' (a special video codec) or 'bursty' as encountered in LAN interconnections; they all go down the same 'pipe'. The aggregate stream is sufficiently fast to carry them all.

An ATM switch is a cell switch; the stream of cells associated with a particular call all have the same header and are processed by the switch purely on the basis of that header. If the switch is fast enough it will appear to be transparent to the incoming cell stream. The promoters of ATM claim that it will provide a single transport medium and a single switch for all foreseeable services. This is what has so excited the telecommunications community: the management problems associated with having to provide separate switches for different services, could be solved at a stroke.

The strategy for ATM can be summarized as follows. Introduce the technology for new broadband services as an overlay; use it to absorb growth in wide area data traffic and to replace other forms of wide area data transport and switching. Then ATM can be used to absorb growth in voice traffic, followed by the replacement of obsolete voice switches.

Yet is this really the Holy Grail, the philosophers stone, the single unifying principle we have all been searching for? To answer this question it might be useful to apply our architectural criteria. Can ATM be provided as an overlay? Yes, undoubtedly. Is the existing transmission infrastructure transparent to the ATM cell stream? Yes. Would an ATM switch absorb the present dominant service; that is voice traffic? Again, is ATM technology; both as a transport medium and a switching node transparent to this dominant voice service? The answer at the present time must be a qualified 'no'. An ATM cell, as currently defined, contains 48 bytes (octets). Thus it takes 48 PCM speech samples to load a cell. But a speech sample is only taken once every frame, that is once every 125 microseconds. Thus an ATM cell carrying a single speech channel takes 6 milliseconds to fill. If this delay is added to the 'padding' delays necessary to remove the jitter caused by the ATM switching process, then it begins to become significant.

Nevertheless, if ATM is applied as an end-to-end, customer-to-customer overlay, this delay is unlikely to affect the perceived speech quality because only one PCM to ATM conversion is required. Trouble starts when a number of such conversions are required in a single connection; for example, when a call traverses a number of networks, not all of which have an ATM capability. This can occur, not only in an international call, but also when regulatory action has fragmented the coherence of a national network.

10.5 OPTICS

First, a brief description of one of the projects in the European RACE programme: it is called CMC or the coherent multichannel project and is aimed at demonstrating how multichannel programme material (for example television) can be distributed by a passive optic network (PON). The same network can also support telephony.

The coherent multichannel system demonstrator is based on the principles of coherent optical transmission and optical heterodyne detection. This represents the optical equivalent of modern day radio transmission based on the super heterodyne receiver principle. The transmitter optical oscillator is a spectrally pure, single mode laser source which can be amplitude, frequency or phase modulated. At the receiver, the received signal is combined with a local oscillator laser whose optical frequency is very closely matched to that of the transmitter. The process of detecting the combined signal and local oscillator laser fields in a PIN photodiode provides a mixing action which generates an electrical intermediate frequency signal which carries the signal modulation. This form of detection offers higher sensitivity than conventional detection methods and the intrinsic frequency selectivity of the heterodyne receiver provided by the IF filter enables a single optical channel to be selected from a large, densely packed optical multiplex, through the use of a tuneable local oscillator laser.

The system, which is shown in Figure 10.2, consists of 10 optical transmitters each transmitting 140 Mbits digital video using FSK modulation. The transmitter units are spaced 10 GHz apart in optical frequency and this spacing is maintained by the multichannel stabilization unit

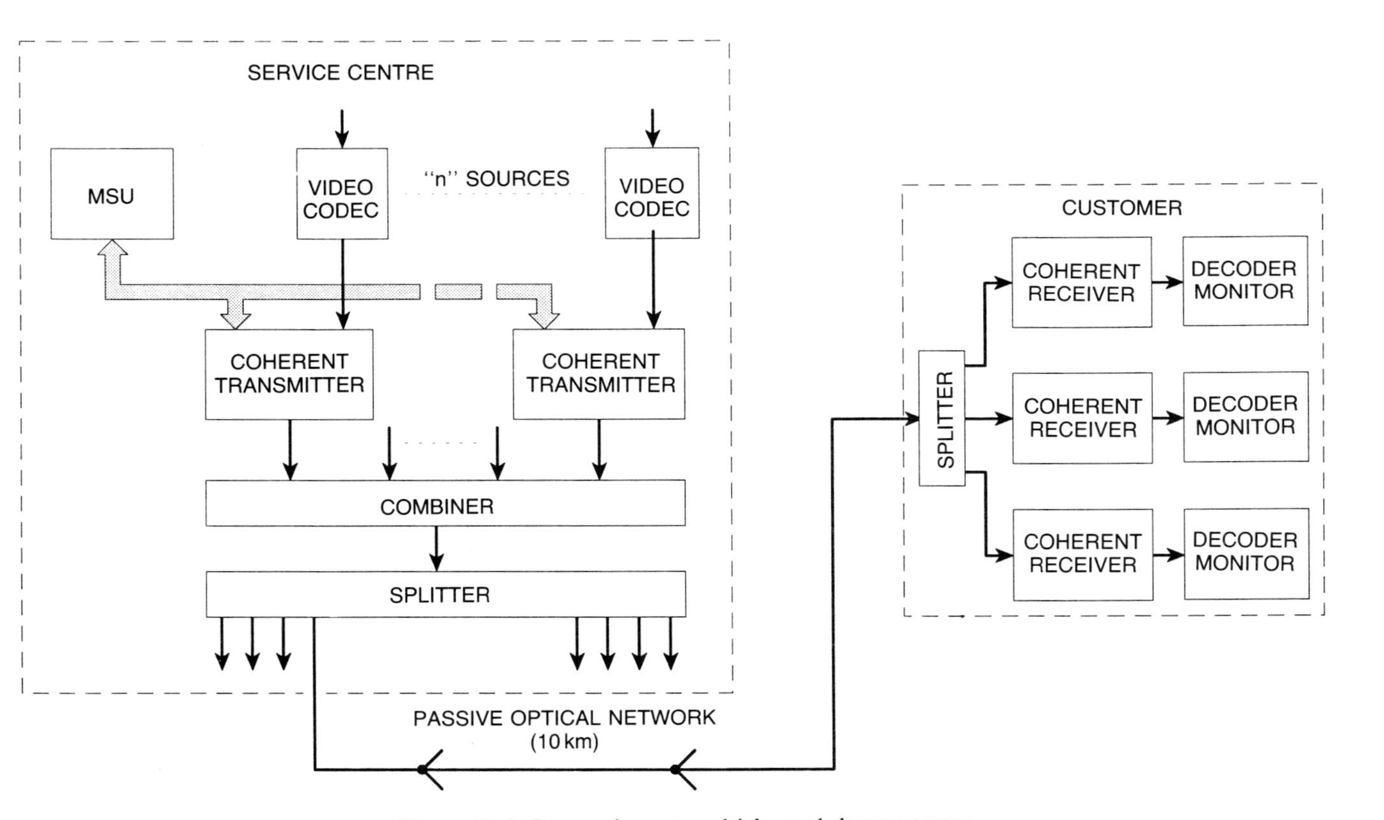

Figure 10.2 Race coherent multichannel demonstrator.

(MSU) which, in addition to initiating automatic start-up of the system, monitors the frequency of the transmitter lasers and implements frequency correction if necessary. These transmitter units are combined and distributed over a passive optic network in which the power budget has been designed to permit up to 1000 terminals to be fed from a single head-end unit. The optical heterodyne receivers use semiconductor local oscillator lasers tuneable over, typically, 200 GHz, though the system requires only 100 GHz.

The principle used in this demonstrator can be generalized into what has been described as 'optical FDM': the creation of a dense optical multiplex, composed of very fine 'lines' or narrow wavelength channels. These channels can be modulated with many different types of signal, either directly or with an external modulator and it is the exploitation of these techniques which will fulfil the real promise of optical transmission: tens or even hundreds of optical channels, each one capable of being modulated with a 10 Gbits digital, or a 4 GHz microwave signal.

Nevertheless, even optical fibres have attenuation and this implies that 'repeaters' will be required to realize a long route. Up to now, these have taken the form of digital regenerators, where an optical signal is demodulated, reshaped and launched into the next length of the fibre. If it were necessary to use digital regenerators for the optical FDM scheme outline above, then the dense optical multiplex would have to be demultiplexed and each of the many hundreds of channels put through a regenerator, relaunched and remultiplexed. Further, the regenerator applied to each channel would have to be specific to the form of the information carried on that channel: as has already been observed, such a regenerator is not transparent.

What is required is the optical equivalent of that mainstay of the old analogue FDM systems: the wideband, linear, low noise amplifier. Optical amplifiers with many of the desirable characteristics are now emerging from the laboratories. The most promising is the fibre amplifier, formed by a length of doped fibre. It will be considered in more detail later.

Just as many customers used to lease copper private circuits – indeed some continue to do so – so potential users are talking about leasing 'dark fibre' on which they can impose whatever signal they wish. If the distance required is within the normal reach of optical signals, say 100 km, this is a perfectly feasible technique and no doubt operators in the future will be prepared to lease optical private circuits in the same way that they now provide the copper equivalent.

However, dark fibres are no use if long distances have to be covered between customer sites and in this case it is possible to predict that customers will lease optical wavelengths, just as they lease a megastream today, and modulate this to suit their specific purposes. Indeed, the development of an optical cross connect where wavelengths are switched instead of digital bit streams can be anticipated. It should be noted, however, that this switch has to be transparent just like the amplifier and it may not be possible to retain the same wavelength if an acceptable level of blocking is to be achieved. Thus we see the requirement for a switch element capable of transposing a narrow optical channel at one wavelength into another with a different wavelength, while preserving transparency – the optical equivalent of the timeslot interchange found in the digital switches of today. It is highly unlikely that such switches will be with us until the next century.

10.6 SATELLITE MOBILE RADIO

As we have observed, satellite links play a major role in today's communications network and, within their bandwidth limitations, provide the perfect end-to-end overlay. Until now, these links have been provided by geostationary satellites and only a limited degree of customer mobility has been possible, as in the Inmarsat network.

Geostationary satellites have a number of limitations, not least the huge expense associated with their construction and launch and now a number of consortia are promoting the use of what are called 'low earth orbit' satellites (LEOs) for a worldwide, portable communication service.

The most ambitious of these projects is called Iridium and has been proposed by Motorola. The system is planned to provide telecommunications services to mobile or static users anywhere on or around the earth; that is on land, sea and in the air. The heart of the system is a constellation of 77 small, low orbit satellites arranged in a number of rings circling the earth in polar orbits, with alternate rings rotating in opposite directions. Each satellite has the ability to illuminate 37 cells, each of 360 nautical miles diameter. The number of cells that are active on any satellite at any one instant depends on the position of the satellite above the earth's surface, that is more cells are required at the equator and progressively less in higher latitudes. Hence the number of active cells and their coverage is continuously changing as the satellites orbit the earth.

A simple block diagram of the system indicating the main components of the iridium radio network is shown, in Figure 10.3. As can be seen, the main components are: the user stations, the up-down links, the satellites, the cross links, the gateway links and the gateways. The simplest type of connection involves a single user station, an up-down link, a single satellite, a gateway link and a gateway which provides the interface to the fixed telecommunication networks and enables communication with

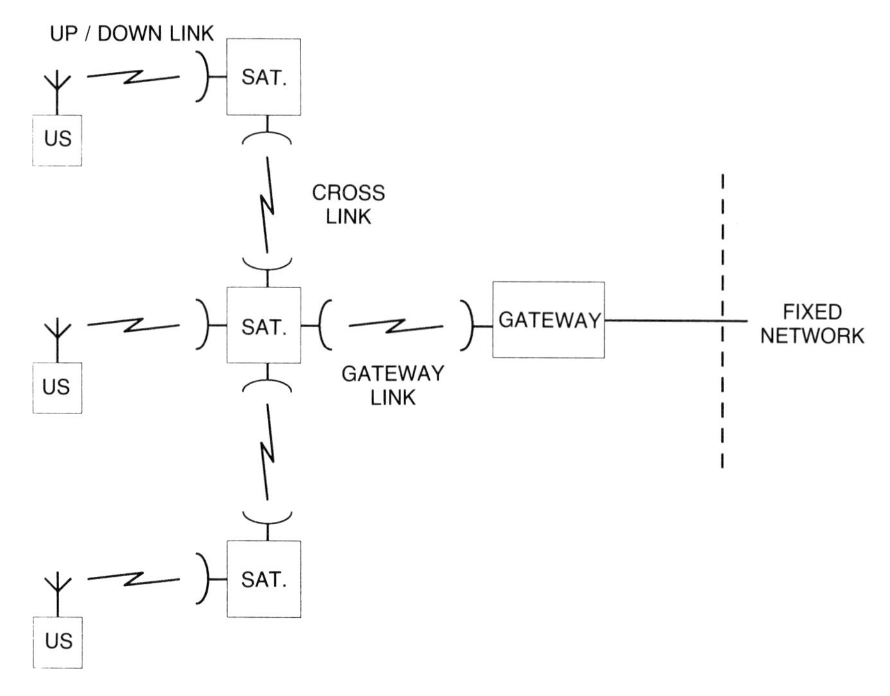

Figure 10.3 The 'iridium' mobile radio network.

anyone that has access to telecommunication facilities anywhere in the world. It should be noted, however, that this simple connection is not as simple as it may appear. As the satellites are continuously orbiting the earth, the radio cell serving a particular user changes approximately every 60 seconds, therefore the connections are continuously being reconfigured.

As can be seen, even from this brief description, this system is highly complex and includes 'on board' switching, as well as the transmission functions normally associated with satellites. Many commentators have put it in the same class as 'Star Wars' in terms of feasibility. Even so, other consortia have proposed similar systems although most do not include on board switching. Nevertheless, if the Iridium type of system is fraught with difficulties and may never achieve its objectives, or indeed see the light of day at all, its aim is perfectly valid. How can a world-wide wireless telephone service be implemented. How can I make and receive telephone calls using the same handset anywhere in the world?

At this stage it is perhaps best to leave the distance domain and turn to the value domain. A number of cellular networks exist, why not make a transceiver which operates with them all? Dual mode transceivers are indeed beginning to emerge. In the USA, there will soon be a transceiver that will operate with AMPS and the emerging TDMA Digital AMPS standard. Advocates of CDMA are advancing proposals for a range of dual standard transceivers to operate with CDMA and AMPS and CDMA and the European GSM standards. Indeed, the Iridium consortium itself has now started to describe its own version of dual standard, portable handsets: giving access to Iridium on the one hand, and either to AMPS or the GSM on the other. However, there is another motive for Iridium activity in this area: it is obvious that the traffic density that can be satisfied by Iridium is limited by its use of LEO satellites.

So it would seem that there would be a reasonable chance, in the early years of the next century that I will be able to make and receive calls anywhere using a single handset. However, whether this is going to be via Iridium, or the new universal mobile telephone service currently being studied by CCIR and ETSI or by using a multi-standard handset, only time will tell.

10.7 ADDING VALUE

While the preceding discussion has concentrated on some of the technologies likely to be found in the distance domain in the early years of the next century, we have also indicated how similar aims in terms of services provided to the user could be achieved by using different technology in the value domain. One example has been the use of DSP techniques to further compress video signals so that they could be carried over multislot connections in the copper network. Another, has been the possibility of using similar DSP technology to develop multi-standard portable radio handsets which adapt themselves to whatever network they find themselves in, to avoid the costs associated with the development of a universal system.

These examples have dealt with terminal equipment, how technology can be applied to terminals to realize services and features which the network may deliver in the future but where that future may be more long-term than can be permitted by the market. I now want to use the example of Intelligent Networks to indicate how value may be added to the network itself. Figure 10.4 gives an outline of the architecture of the intelligent network. The various services and service features associated with the IN are implemented by what are called service logic programs (SLPs) which are located in what is called the service control point (SCP). This SCP is a centralized computer with a large database. It is connected to the switches it serves by an enhanced version of the signalling system (C7) used to control ordinary telephone calls. The number of SCPs in any one network depends on the data that has to be stored, the number of

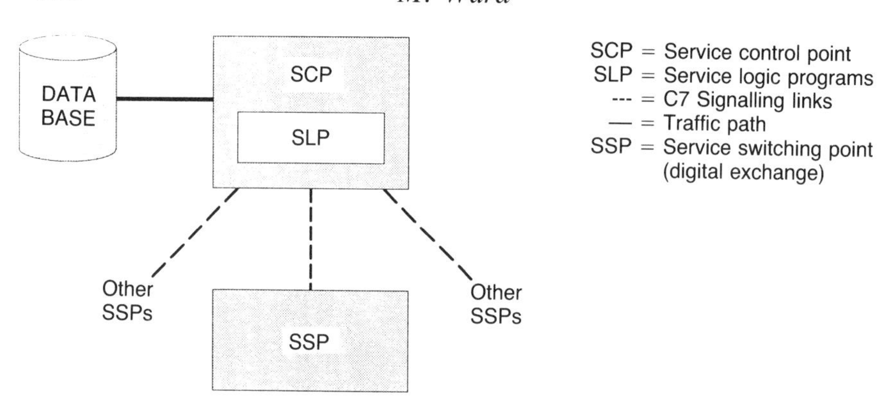

Figure 10.4 Outline architecture of intelligent networks.

accesses (transactions) that have to be processed and how often the data has to be modified.

A typical application of IN is the 0800, or Freephone service. When a customer dials 0800 . . . , the switch triggers an access to the SCP. It is important to note that the call is not switched to the SCP – no traffic path is established – only control is transferred using messages passed over the C7 signalling system. A service logic program in the SCP examines the incoming data – what the customer dialled, and where he dialled it from – adds 'centralized' data – such as the time of day, date, etc. – and uses this combination to look up the database. The result of this look-up will tell the SLP how to route and who to charge for the call. Relevant parts of this information are then passed back to the switch and the call is established.

Call screening provides another example. Here IN techniques are used to ascertain whether customers have a right to access the services they have asked for: to authenticate the requests, to verify that a customer is the person he says he is. At the moment identification is provided by the customer in the form of a pin number and this pin is used in the database look-up process. In the future security can be increased by the use of voice recognition techniques to verify the speaker. Indeed, by the turn of the century it may be possible to do away with the keypad altogether and just speak into the telephone.

What has been described as 'personal mobility' can also be implemented by IN techniques. In this case, a customer can personalize any telephone either by using a special code, a charge or credit card, or a smart card and turn that phone into his own. Not only will he be able to make calls from the phone, charging the calls to a single account, as in today's charge card services, he will be able to use his own special profile of features and he will be able to have calls routed to him on that phone as though it were his own. Security screening could operate as described above.

Early versions of this service will soon be available on the European GSM cellular network. It will be possible, indeed necessary in some cases, to personalize a GSM hand portable or mobile phone by inserting a special sort of smart card. This will allow a traveller to enjoy the benefits of the Pan European Digital Cellular Network (GSM) without having to carry a hand portable with him. He will be able to rent a new phone when arriving at his destination and personalize it using his smart card. All calls will be charged to the traveller's account and calls destined for him will be routed to him wherever he is.

An extension of the personal mobility is 'personal numbering' where everyone will have a personal number, for example, issued at birth, rather like a National Insurance number.

Now a number of actors in the telecommunications play find the capability of INs very attractive. Independent service providers see the possibility of INs installed by the network operator as a means of reaching universal audiences from a single point. The operators are not blind to these possibilities either, but they also see that INs can be used to deploy new services rapidly and to achieve independence from their switch vendors. They can design new services themselves, or employ an independent third party to do it for them and deploy the new service from a single point without having to modify all the switches. However, before this aim can be achieved modifications have to be made to all the switches to realize the standard interface between the switch and the SCP which is required for vendor independence. Thus some of the aims are mutually contradictory, at least in the short term.

IN has been introduced as an add-on or as a value-added service to the existing network; as a way of bringing new services to anyone who has a telephone. The techniques employed can be extended to third party services such as home banking. They can also be employed to improve the functionality of the network itself. Thus the technology, which has been applied, initially, in the value domain could be embodied in or absorbed into the network itself.

The initial IN services described above are simple examples of distributed processing. The term refers to a situation where, the client, performing a certain task finds that it cannot complete that task without help from another process, so the client sends a message or calls the other process, which we shall call the server, to request its help. In the IN examples we have given, the client will be in the switch and the server will be one or more service logic programs in the SCP.

This principle can be generalized to the point where there are many clients and many servers and where the roles can be reversed. Thus we can see the possibility of removing all but the most basic functionality from the switches and putting all the processing associated with features and all the data related to customers, in separate processes running on remote computers. These remote processes (or objects) need not run on hardware belonging to the network operator; for example, they could be part of a PABX or contribute to the service offered by a third party service provider. They could be provided once for the whole network if their role was to implement a new or rarely used feature or replicated many times if their functions were required by many customers or for every call.

There are many who would claim that this distributed processing actually simplifies the process of software engineering, as well as allowing the rapid introduction of new features, but we don't have the space to discuss that here.

10.8 APPLICATIONS OF THE TECHNOLOGY: SOME POSSIBLE NEW SERVICES

The Americans have coined a very descriptive phrase for one type of service made possible by broadband transmission and switching: they call it 'video dialtone'. By this they mean all types of video services on demand, services that are as easy to access as picking up the telephone. . . . And it need not be just one service, it can be a mixture of image, text and voice. For example, dialling a movie either from this weeks selection or from a whole library, with no more delay than it takes to load the video jukebox, unless all the copies are in use. But even this could be solved by a multi-headed disc. Then what about the newspaper with the still pictures replaced by short videos, or the shopping catalogue where the photo of a dress could become a movie and you could actually see how something worked. Another example would be a moving maintenance manual; if your car broke down you could dial up a service which could show you how to fix it in a video sequence which could be frozen at crucial points in the process.

It is in the field of education, however, that many think such multimedia services will achieve their first breakthrough. An Open University course where the student can see the tutor and vice versa, illustrations can be visual and animated, and where experiments can actually be demonstrated is now becoming possible.

Nevertheless there will be many who will protest that such services are beginning to be available now; why wait until 2005? However, this is to miss the point because such services won't make much impact until they can be used by 5% of those who now own a telephone and I would estimate that it would take at least ten years to achieve that sort of penetration.

But how much bandwidth will such services need by then? After all, very acceptable video quality can be achieved with 2 Mbits even now. Indeed, how much bandwidth will HDTV require by 2005?

While multimedia, as described above, will have some application in the business world, collaborative working using multimedia capability is likely to make the biggest impact. To illustrate this consider three engineers on different sites, working on the same design, using the most modern CAD equipment. They will need to pass high resolution CAD images to each other and work on them together, while accessing other information from databases and perhaps performing simulations on remote super computers. This would seem to be an ideal application for ATM: high bandwidth would be required to transfer the complete high resolution images but their manipulation would require a comparatively trivial data rate. Nevertheless, it will not be the bandwidth which will cause the difficulties, but the control signalling, the commands, the man-machine interaction. A single specific application on a private network might be relatively easy to set up but a general purpose, collaborative working, service on the public network is another matter altogether.

Taking an example, now, from the value domain, voice processing is likely to make the breakthroughs in the next ten years that have alluded it in the last twenty years. Machines capable of recognizing continuous speech with a 20 000 word vocabulary on a speaker independent basis ought to be possible by the end of this century. Of equal importance is the ability to translate text into speech in a realistic manner. Language translation may also be another service which could be provided over the public network. Speaker authentication, that is verification that the speaker is who he says he is, is just emerging from the laboratories now and ought to be commonplace by the next century. Speaker recognition, that is picking out an individual from a crowd is much more difficult and in any case is likely to be a much more specialized service.

The final application for a future broadband service considered in this paper is ‘virtual reality’, which can take two forms. The first involves a 3D image of the ‘virtual world’ on an otherwise ordinary workstation together with a more advanced version of the standard mouse to allow the user to interact with this world.

Realistic sound could also be provided. The second is more ‘inclusive’ and attempts to intimately involve the ‘inhabitant’ of the virtual world in what is going on. The inhabitant can replace the workstation screen by stereoscopic goggles and the view is changed when he turns his head. He can point at a virtual object and gloves can be provided which give some sensation of feeling and pressure; however the inhabitant still has to navigate through his virtual world using a 'space ball' or some other advanced version of a mouse.

While, at the present time most of the tentative applications of virtual reality are 'stand-alone' and have no influence on, or use for telecommunications, this will not always be the case. For example, let us consider a future estate agent where all the dwellings on offer would be modelled in a virtual world and a prospective buyer would be able to walk round them. Most descriptions of this type of application have assumed that the buyer would do this in the estate agent's office, but suppose he could do it in the comfort of his own home and the estate agent could 'accompany' him round the house. This would involve 'remote inhabitants' sharing the same world and would have many implications for telecommunications. If we are going beyond 3D representation of objects on a flat screen or stereoscopic goggles, to 'real' 3D objects, say, in the form of holograms, the demands on telecommunications would obviously increase. As another possible application let us consider a design team in one country wishing to show the results of their labours to a marketing team in another. Virtual reality with 3D 'objects' such as holograms would allow the marketing people to walk round the prototype and feel its surfaces, etc.

However, although applications such as those described above would provide interesting and possibly useful applications of virtual reality, they are unlikely to pull though a massive deployment of broadband capability to the residential customer. Virtual reality games on the other hand might just do this. They might provide that element of obsession, and not just with children, which will be necessary to generate the intensity of demand required to justify the expense. As an example, consider one of the Nintendo games for their Game Boy product. It is called 'Golf' and I have watched adults become absorbed in this to the exclusion of everything else. A virtual reality version of this is fairly easy to envisage but in order to provide the pull for broadband communications, players must want to enter their virtual golf world from dispersed locations. It is not enough to meet in the bar around a common system.

And then there is the virtual world inhabited by the users of the chatlines and the Minitel 'services' that will become so popular in the next century. Remember the 'Feelies' in Aldous Huxley's *Brave New World*. There the audience had to go to a cinema to join in this experience, in the next century we may be able to do this in the comfort of our own homes.

11

Mobile communications in the 21st century

Prof. R. Steele, Electronics and Computer Science Department, University of Southampton, and Multiple Access Communications

11.1 INTRODUCTION

If a scientist in 1892 had been sufficiently foolish to predict the development of communications in the 20th century, we can be certain that his prophesies would have been substantially incorrect. He perhaps would have forecasted correctly that the telephone would be nearly ubiquitous, but it is doubtful that he would have concluded that digital communications, with its roots in telegraphy, would be used to convey the human voice. Marconi had not conducted his famous radio communication experiments, but Hertz had demonstrated radio waves. It was therefore predictable, given some unforeseen and clever means of production and detection of radio waves, that these waves could be transmitted around the world. Satellite communications would have seemed unthinkable in an age before the South Pole was discovered, the first car appeared, or the first flight was made.

The communications prophet in 1892 might have taken the view that as communications currently involved the transmission of messages (telegraphy) and speech (telephony), they would eventually be extended to the transmission of the other human senses. Transmission of images over vast distances would have been predictable, but the means to do it would have been enormously beyond his scientific knowledge. It is likely that he would not have mentioned such deliberations as he might have appeared to be more like an astrologer than a scientist. He would also know that the development of both communications and society were intertwined, and that in order to predict the future of communications he would be

Communications After AD2000. Edited by D.E.N. Davies, C. Hilsum and A.W. Rudge. Published in 1993 by Chapman & Hall, London, for The Royal Society. ISBN 0 412 49550 3

well advised to consult the works of people like H.G. Wells. What he could not have predicted was the transistor, or the deployment of millions of them on a single chip, optical fibre communications, the phenomenal computer power that developed in the 20th century, information theory, the massive and complex aircraft and missile industry with its electronics and communications, the bull's-eye accuracy of the military machine, robotics, and so forth; a list that has depended on discoveries in the basic sciences. The communications networks in the 20th century are complex, with their reliance on radio, cable, fibre, computers, microelectronics, etc., and it is these networks that are essential to the day-to-day running of society.

There now exists a considerable theory relating to prediction. Indeed predictors are embedded in our speech and video codecs, in our missile systems, in handling fractures in metals and in weather forecasting, to mention just a few of their applications. In more complicated scenarios, such as earthquakes and the national and international economies, prediction is found to be poor. Predicting what will happen in communications in the 21st century is far too complex to have any chance of success, and the situation is worse when it comes to the topic of this talk, which is the prediction of mobile communications.

However, all predictors utilize known events that seem to be correlated with the event to be predicted. Thus we may be able to predict the type of mobile communications that will exist during the first one or two decades of the 21st century. These predictions will contain certain errors, particularly if there is a 'breakthrough' in material sciences. Predictions farther in the future become increasingly valueless. Thus our predictions well into the 21st century will not be based on science, but on hunches. Faced with an impossible task we will proceed on the basis that what follows may be thought provoking.

11.2 THE CURRENT SCENE IN MOBILE COMMUNICATIONS

The 1980s laid the foundations of public mobile communications. During this decade, so-called analogue cellular mobile radio networks, such as the Nordic Mobile Telephone (NMT), the American Advanced Mobile Phone Service (AMPS), the British Total Access Communications System (TACS), and the Nippon Advanced Mobile Telephone System (NAMTS), were deployed. They all have similar features (Jakes, 1974). Key to their success is the notion of cells, and hence the term 'cellular radio'. A cell is a geographical area containing a base station (BS) which communicates with mobile stations (MSs) in its cell using UHF radio signals. The frequency band allotted by the regulatory body is divided up amongst a cluster of cells. Each MS transmits to its BS using an assigned channel which occupies a relatively narrow frequency band, typically 25 kHz. Every MS within the cell cluster uses a different frequency band. However, the number of mobile users that can be supported by the cluster is relatively small, typically less than one thousand. In order to dramatically increase the number of users, clusters are tessellated, with each cluster re-using the same allotted spectrum and supporting the same number of users. Consequently there are MSs in each cluster using the same frequencies. The resulting cochannel interference is kept to acceptable levels by the physical separation between cochannel BS sites.

All these first generation cellular radio systems use frequency division multiple access (FDMA) and have separate bands for down-link transmissions from BS to MSs and up-link transmissions from MSs to BS. Speech transmissions are the main form of traffic and are conveyed using frequency modulation (FM). Because FM radio is used, the systems are said to be 'analogue', but this term is misleading as all the network control is digital. Indeed, the complexity in these systems lies in the digital transmissions and signal processing required for registering calls, effecting handover of calls as the MSs travel between cells, terminating calls, controlling the transmitted power to decrease cochannel interference, interconnecting to the PSTN, and so forth. The analogue FM speech transmissions are comparatively trivial.

By the early 1980s there were serious proposals for second generation mobile systems based on all-digital techniques. A conceptually advanced proposal, known as the CD900 system from SEL, appears to have been the catalyst that eventually had the Europeans working together to define a pan-European cellular system. By late 1986 a trial was held in Paris where eight contending systems were compared. From that trial, and after much technical committee work, a system emerged known as GSM after the Groupe Speciale Mobile committee that finalized its specification. By July 1991 the system, now called the Global System of Mobile (GSM) communications, began to be installed.

Meanwhile the Americans, suffering from a dearth of mobile channels, decided to introduce a digital system called IS 54, where each 30 kHz AMPS channel is replaced by three digital channels organized in a TDMA format. The Japanese have essentially followed the same path. These systems will be deployed in the early 1990s.

These second generation cellular systems are flawed, because although they digitally encode speech and transmit it using binary modulation, they use the large cell philosophy of the first generation analogue systems. By changing to TDMA they significantly increase the transmission rate which causes intersymbol interference. While TDMA offers significant advantages in the radio frequency (RF) design, complex baseband signal processing is required. The use of large cells also requires high radiation power levels and this, along with the signal processing, requires the batteries to be frequently re-charged. Both GSM and its sister, the digital communications system at 1800 MHz, known as DCS 1800, are essentially large cell systems designed for national coverage, where they will have to operate in both large rural cells and small urban cells. Although the most efficient way to increase spectral efficiency is to reduce the cell size, the hand-held transceivers designed for GSM and DCS 1800 when operating in very small cells, known as microcells (Steele, 1985; 1989), are unnecessarily bulky and complex (Steele, 1990).

Current cellular radio is locked in a Marconi time warp, often blasting large amounts of electromagnetic energy from large BSs whose antennas are located on the top of tall buildings. Although the second generation's BS equipment is significantly smaller than the room-size first generation equipment, it is nevertheless of cupboard size. The network designer must still consider the number of BSs to be deployed because of their cost, difficulties in obtaining suitable sites, and the expense of connecting them into the network. Later we will argue for small cells and inexpensive BSs and infrastructure costs. The frequency bands currently used for cellular radio are limited, around 950 MHz and 1.8 GHz. The power levels can be very large (up to 300 watts in a GSM BS).

During the eighties work was started on cordless telecommunications (CT). The concept was to produce consumer-type voice communicators having limited range for use in office environments. The first digital CT, called CT2, was deployed at the end of the eighties amid confusion. Using a single channel per carrier in a time division duplex (TDD) mode, it will be joined in the nineties by the 12 channels per carrier TDD system known as the Digital European Cordless Telecommunications (DECT) system. Because CT2 and DECT use microcells their hardware is relatively simple and inexpensive. Their bit rates per channel are relatively low (32 kb s^{-1}), although the bit rate per carrier in DECT is 1.152 Mb s^{-1}. We may expect CTs to cover both buildings and short segments of streets and be serious competitors to digital cellular radio in urban and suburban areas, particularly as the country becomes increasingly covered by small microcellular BSs located on ceilings, outside walls and lamp posts. Currently the radiated power from CTs is limited by regulation, restricting their range to below 200 m.

By the late eighties Qualcomm Inc. in the USA was advocating the use of code division multiple access (CDMA) for cellular radio. At the time of writing they are attempting to get their CDMA system adopted as a standard in the USA and, if successful, it will compete with the technically inferior IS 54 system. Opting for CDMA was ambitious because the radiated power from each MS must be tightly controlled as the number of users the system can support is determined by the ability to ensure that the received power at the BS from each MS is within 1 to 2 dBs. As duplex bands are used, and large and independent fades occur on each band, this power control must be exercised some 800 times per second. However, Qualcomm have been able to achieve this control, and to introduce soft handovers between BSs and between cell sectors, employ single cell clusters and use path diversity. This system is conceptually more advanced than any other, but the complexity is less than GSM in that its infrastructure costs are lower and the portables no more complex. It is based on a sound theoretical foundation, and we may anticipate the further development of cellular CDMA, now that the initial problems have been understood and solved.

The current and proposed second generation mobile communications systems, both cellular and CT, are digital, mainly convey voice but with some low data rate services, inter-work with the PSDN and, in some cases, are compatible with limited ISDN requirements. Our mobile radio networks are small adjuncts to our massive fixed networks. Most traffic to and from mobiles originates and terminates on fixed terminals. These terminals are numbered, not the people who use them.

It was during the eighties that CCIR formed a committee to study a future public land mobile telecommunication system (FPLMTS). Their brief is to examine how to provide a variety of services to mobile users, for a wide range of user densities and over most parts of the world. The committee continues to deliberate, although FPLMTS is scheduled to start operating around the year 2000. FPLMTS will provide radio access to a wide range of telecommunication services interworking with other mobile and fixed networks, and will be ISDN compatible. The services will include telephone, audio, message handling, teletex, paging, telefax, point-to-multipoint, data, videotex, video telephone, location of user and multimedia. FPLMTS has opted for signalling interfaces according to the open system interconnection (OSI) model, end-to-end encryption, the use of mobile-satellite communications to users in remote areas, world-wide roaming, and so forth. To achieve these ends it has proposed radio interfaces between personal stations (PSs), carried by pedestrians, and their cell sites, MSs (vehicular) and their BSs, satellite repeater and the mobile earth station (MES), and paging PSs and their pager base stations. Attempts will be made to make the architecture of FPLMTS independent of technology, and to use, where applicable, existing international standards, e.g., CCITT signalling No 7.

In Europe there is the universal mobile telecommunication system (UMTS) project with aims less grand than those of FPLMTS, but with, perhaps, a more specific set of goals that are achievable by the 21st century. For example, UMTS is concerned with the type of radio access, error coding strategies, modulation, equalizers, handover algorithms, dynamic channel allocation, low delay low bit rate high quality speech codecs, packetization, personal numbering, integration with the integrated broadband communications network (IBCN); the list is very long. The results of the UMTS project are available to the FPLMTS committee.

11.3 PREDICTIONS FOR MOBILE COMMUNICATIONS BY THE YEAR 2020

We will refrain from predicting mobile communications in the so-called under-developed countries, as by the year 2020 they are unlikely to have reached the current levels in the developed areas. For example, many villages in India do not have a single telephone channel, so rectifying this situation will take a long time. Nor will we predict the military use of mobile communications, which is closely related to the relationship between nation states and the changing authority of the United Nations. Our predictions will be confined to public mobile communications on a global basis, but with the emphasis on the use of these systems in developed countries. We have selected three arbitrary dates in the 21st century for predicting what might occur around these times. The timing of our predictions will be inaccurate, but of greater concern is whether they will occur at all! However, we will continue as if blessed with perfect foresight.

The process, already started and therefore predictable, of deploying communication networks that will enable people to communicate at any time, at any speed and from any location, subject to teletraffic demand and the users' willing participation, will be partially established by 2020. In many networks users will be able to electronically personalize a terminal. This terminal may be owned by the user, tethered, or someone else's terminal commandeered for the duration of the call. The terminal may range from being complex and large, to a light-weight hand-held or wrist-watch size small communicator. This concept of personalizing a terminal has already been proposed and adopted by FPLMTS. It is known as universal personal telecommunications (UPT), and will enable users to dial a personal number, rather than the number of a terminal. By this means a call will be diverted to the terminal that is closest to the user and to which he has accessibility. Users will be assigned a personal number, and it is this number which will be dialled by others, or used by the subscriber to personalize a terminal. UPT will allow users to decide which calls to accept, divert or ignore. Speech authentication and verification will be used by UPT. Billing will be against the user's number, and the decision as to who pays for the call will be resolved.

Towards the end of the 20th century all CT handsets operating from telepoints will be given pagers, and it will be rare to find a basic one-way pager service. With the increase in the number of telepoints, CT microcellular systems will appear in city centres. These CT street microcellular systems will merge with the office CT systems by the early 21st century to give continuous roaming over urban conurbations. The CT handsets will still be lighter, cheaper and with longer battery life than the cellular portables. The operators of conventional cellular networks will respond with portables, resembling CT handsets in many ways, that are especially designed for city use. As all cities will have only microcells with larger, oversailing macrocells, the interworking of CT networks with cellular systems will be replaced by one type of network which will allow a lightweight portable to be used nationally, as all the cells by 2020 will be no larger than 2 km in the UK. The concepts of CT and cellular will therefore blur, the systems merge and simply be called, 'mobile networks'.

With the microcellular systems will come small size BSs radiating low power (microwatts to 20 mW) below the urban skyline in cities, and in general not higher than 15 m elsewhere, to mobiles which are in close proximity. Office blocks will be 'wired' with optical fibres to facilitate the deployment of microcellular clusters. For example, there may be many microcells per floor, or one per floor, or one per office, and in general a mixture that is dependent on the prevailing teletraffic demand. Microcell BSs located on the outside walls of buildings, on lamp posts, distributed around parks and football stadia will be connected via fixed links to the network. There will be different and overlaying microcellular clusters for vehicular and pedestrian MSs. Mobile radio transmissions will be to handheld portables over short distances, often less than 100 m. This intense deployment of BSs will be throughout many cities and towns and along roads. It will be made possible by transmitting microwave signals in an optical format over fibres, or via a higher frequency (20–90 GHz) carrier, reducing the BS to a relatively inexpensive item.

As the BSs and the means of interconnecting them will be inexpensive there will be a significant change in our philosophy of deploying BSs. Currently, an operator must calculate the cost in deploying a BS, and the channel utilization he can expect. The notion of deploying a BS that is rarely used is appalling to an operator in 1990. The attitude will be totally different in 2020. In the same way that we install light bulbs to provide light when we need it, but are not concerned if they are rarely switched on, so we will deploy inexpensive shoe-box to match-box size BSs.

Networks will have distribution points which will transmit and receive via radio or optics large amounts of data over short distances, like an upmarket telepoint, where a user is paged to receive, or goes to receive, 20–1000 $\mathrm{Mb\,s^{-1}}$ of data delivered on carriers greater than 50 GHz. These distribution points will be used for special services.

Even by 2020 the networks will still be dominated by speech transmissions. These will still be narrow band telephony, but wider bandwidth speech at 7.5 and 15 KHz will be increasingly transmitted. Speech codecs will probably operate from 1.2 Kb s^{-1} to 64 Kb s^{-1}. Image transmissions will be a growing market for mobile. Mobile video conferencing can be expected to operate over the range from 8 Kb s^{-1} to 64 Kb s^{-1}, while mobile broadcast quality high definition television will be available at rates from 0.5 up to 10 Mb s^{-1}, depending on the type of image presentation. Fax and slow-scan television will be common place, as will be mobile 'virtual world' communications. Mobile computer data transmissions will be available at a wide range of rates up to 20 Mb s^{-1} per user, and transmitted where possible in unique bands that use large block interleavers, powerful concatenated channel codecs and ARQ.

The choice of modulation and multiple access methods will be independent of the complexity of signal processing. Due to progress in microelectronics, advances in battery technology and the use of microcells with their low radiated power levels, formidable signal processing will be used in hand portables. Solar charging of batteries will be a common feature. There will still be vestiges of TDMA and binary modulation, but adaptive TDMA with multilevel modulation, such as quadrature amplitude modulation (QAM), will be used. CDMA will grow in acceptance as the multiple access method for mobile communications, although its deployment rate will be delayed by the existence of other multiple access methods and entrenched non-mobile services.

Satellite mobile services will be widely deployed where terrestrial infrastructure costs are prohibitive and the number of users small. In developed countries satellite mobile communications will be used in aeronautical, maritime, and wilderness environments. Low altitude orbital satellites using spread spectrum will provide basic speech and data services.

The networks with their synchronous digital hierarchies (SDH) will increasingly employ packet communications for all services, and adaptive multiple access methods with statistical multiplexing will be preferred. Virtual circuits will be less popular than routeing packets with cloning. The debate as to how much intelligence to give the network infrastructure relative to the packets will be unresolved. ISDN will be used for many mobile services. SDH will be transmitted to mobiles on some services. The OSI model will be challenged as too cumbersome. The billing will be registered in each network as the packet passes through the gateways.

There will be moves to get WARC, or its then equivalent, to remove all terrestrial non-mobile radio users from the spectrum between 50 MHz and 10 GHz. By the year 2020 we may anticipate that mobile communications will occur in specific bands between 0.8 and 3 GHz. Different services on the same radio bands will still be done, but some bands will be service specific. For example, some services requiring high bit rates, low delay and low BERs will be on different bands from those conveying fax, slow scan TV or speech signals. To clear the bands for mobile use will be difficult because of existing services and self-interests. However, as mobile communications can only operate in a restricted range of frequencies they must be given priority. This will mean that Government will have to tackle the strong satellite and broadcasting establishments.

In some ways it is unfortunate that mobile networks have different owners from fixed network operators, as the two networks are intrinsically linked. Given the deployment of an increasing number of cellular and CT network operators, we may expect cellular and CT systems to grow into complete national networks, communicating with people in their homes via radio. The fixed network operators, like BT and Mercury, will introduce fibres into many homes, providing a dramatic increase in the range of services. This will force the mobile operators to do likewise. Once the fixed fibre networks become ubiquitous they will have mobile radio distribution points that are essentially small BSs which are located in buildings or mounted on outside walls. Mobile transmissions to users in the street will be more frequently from BSs located in buildings rather than from street located BSs. However, these numerous national or part-national networks will have to interwork, and the complexity in the call routeing and charging will be great.

11.4 PREDICTIONS FOR MOBILE COMMUNICATIONS BY THE YEAR 2050

The fibre network will have grown in complexity and density throughout the century, and by 2050 it will be ubiquitous in all areas of the UK. It will be interfaced with low power, short distance radio to provide almost complete mobility for users. From their homes people will have access to huge data banks and from them a vast variety of entertainment and educational services. The result will be the demise of the broadcasting authorities who will have mutated to programme production companies, inserting their products into the data banks from where they can be extracted by home users. Their broadcast bands will be used for mobile services, with the consequence that the mobile band will stretch from 50 MHz to 10 GHz, with sub-bands that will be service specific. However, television braodcasting of many channels will still occur in bands above 40 GHz. Satellites will be used in a role that suits them, namely, broadcasting, where the propagation delays are not noticed.

A century will have elapsed since Claude Shannon did his monumental work that led to the establishment of Information Theory. His great mathematical theory (Shannon, 1948) was difficult to grasp and seemed irrelevant to most communication engineers during the second half of the 20th century. During this period engineers designed equipment constrained by many factors: cost of active devices, complexity of design, battery life, size and weight of portables, network infrastructure costs, etc. Much time was spent designing robust speech and image codecs to avoid the use of channel codecs. Spectrum efficiency had a restricted meaning, often being equated with the bandwidth occupancy of the modulated signal. The idea that spectral efficiency was intrinsically linked to multiple access methods, cellular patterns, network design, blocking probabilities, and so forth, only began to be appreciated by the end of the 20th century.

The essential factor that will enable Shannon's thesis to be implemented will be the exponential advances in technology. Microelectronics coupled with photonics and technologies to be invented will overcome the complexity in equipment design, BSs will be match-box size, inexpensive and ubiquitous due to radio conveyed over fibre and also radio conveyed by radio, and by 2050 the battery problem will be solved. Speech codecs will operate at bit rates below $200\,\mathrm{b\,s^{-1}}$ as predicted (Haskell and Steele, 1981), but will suffer from delay. Higher bit rates of about $1\,\mathrm{Kb\,s^{-1}}$ will be preferred as the delay will be acceptable for conversational speech. Video codecs for mobile video telephony will operate from $4\,\mathrm{Kb\,s^{-1}}$ to $4\,\mathrm{Mb\,s^{-1}}$ depending on picture quality. The bit rates for the speech and video codecs will be variable. As advocated by Shannon, the vulnerability of the speech and image codecs to errors will not be an issue is designing these source codecs. They will be designed so that the entropy of the speech and images are virtually equal to the entropies of the encoded signals, which should be synonymous with the minimum bit rate for imperceptible degradation. The source codec bit rate will be increased using forward error correction (FEC) coding optimized for both a variable source rate and a Gaussian channel.

The FEC data will be interleaved and spread using long chip codes over a wide band of frequencies and have the characteristics of Gaussian noise. Multiple access will be based on the principle of CDMA as succinctly argued by Viterbi (1991). Complex power control will be essential to ensure the control of cochannel interference. As all the signals are like wideband Gaussian noise, the despreading process will mean that receivers will only have to deal with the fraction of the noise that occurs in-band. Shannon's notion of not discarding information until the final decision is made, e.g., as in soft decoding, will be embodied both at the physical layer and in the network protocols.

Although most mobile radio signals by 2050 will be of the spread spectrum type, a few narrow band signals will still exist. However, these two types of signals can co-exist in the same bands. The immensely more dense communications infrastructure will mean that, except for special cases, all transmissions will be at power levels tens of dBs below those currently used. The crude PMR networks will have been replaced by ones that use cells of sizes compatible with the public mobile networks. The Rambo concept of blasting large amounts of electromagnetic power will have been discarded, except in all but isolated communities. Power levels will be harmonized to allow the widespread use of spread spectrum techniques which will facilitate different services in the same frequency band. High quality mobile networks will result.

The submarine cables under the oceans will be connected to floating unmanned service vessels held in position by power extracted from wave motion. These vessels, less than 500 km apart and linked by optical fibre submarine cables, will communicate with aircraft flying in communication cells ranging from 6 to 16 km elevation. They will deliver and receive 50 Mb s^{-1} to any aircraft using spread spectrum techniques. These marine based radio distribution points will have a superior performance to satellites as they will have much greater teletraffic capacity and smaller propagation delays. The vessels will also provide communications to passing ships and submarines. The aircraft and ships will be moving microcells with mobile users able to communicate in a comprehensive way to any location in the world.

On land, public transport in the form of buses and trains will be moving cells that will embrace the vehicle plus its immediate environment. As a high speed train is also a high speed cell, it will inflict high transitory teletraffic loading on each part of the network as it travels through it. Similar comments apply for motorway traffic where many cars will also be BSs for other cars. The network must therefore be able to accommodate 'solitons of teletraffic' in its highways.

By 2050 the use of satellites for mobile communications will be in decline. Low orbital satellite systems used in the beginning of the century to provide communications to shipping, aircraft and users in remote locations, will be phased out as un-manned mobile radio communication sky-platforms are increasingly deployed. There will be two types of platforms. One held in geostationary orbit, of vast dimensions able to provide high bit rate communications via spot beams whose surface illumination will be of city size. The other type of platform will be tethered to the earth and located up to 30 km in height and placed between the aircraft flying lanes. Barely visible from the earth they will be able to deliver many services. They will be held on station by power conveyed to them via their tielines, and these lines will also house the fibres that convey the teletraffic with the network. Alternatively these platforms could be untethered, hovering, and therefore capable of being rapidly redeployed. The hovering platforms would communicate to earth via radio. The tethered or hovering platforms will be able to track 'solitons of teletraffic', rather than force the task on the terrestrial networks. For example, the platforms could handle the teletraffic from high speed trains, highways, aircraft and ships. They can be rapidly deployed when disasters occur, for example, the rapid provision of communications to a city which has been devastated by an earthquake.

11.5 PREDICTIONS FOR MOBILE COMMUNICATIONS BY THE END OF THE 21ST CENTURY

After a century of change, the tentative steps in the 1990s to bring some sanity to motor transport will yield a vehicular transportation system where cars move in convoys along highways, with each car behaving like a small railway carriage uniquely designed for a user or a family. The cars speed and lane changes will, in general, be controlled by a complex mobile command and communication system. Unlike railway carriages that are coupled to make a particular train, when the car leaves the highway for another road, it joins a new road-train, but still under the control of the transportation system. The driver's skill will only be required in limited and benign situations. The mobile transmission rates to each vehicle will be many $\mathrm{Mb\,s^{-1}}$ and delivered over short distances. The vehicles will be electrically powered. Route guidance will be automatic and vehicles will have become moving communication systems, handling a wide range of services. It will be rare to use conventional voice telephony, as it will be replaced by video telephony.

The world by the end of the 21st century will not consume significantly more energy than it does today. The communications revolution will have had a dramatic effect on society. No longer will vast numbers of people travel many miles each day to work in urban conurbations. Instead they will work at home or in local centres. The saving in energy consumption will be dramatic as will be the diminution in pollution. Travel will mainly be for leisure and pleasure. The distribution of the population will be dispersed, with the minority of the people living in large cities. It is indeed fortunate that there is more silicon in the world than oil, and that its supply is endless for our needs. People will be able to communicate at terminals anywhere in the world, and they will always have easy and convenient access to the networks.

Although the reduction in the use of fossil fuels will result in less pollution, the exploitation of seas will be underway. The communication network in place by 2050 for communication to aircraft and ships will now be assisting fish farming and mineral extraction. Oceanographic hydro-electric schemes will abound. Efficient mobile sonar communications will be in use to high speed submersibles, personnel housed on the ocean beds, and to cargo submarines.

Cybernetic factories and intelligent buildings will use mobile communications for their machines and robots. Aviation will have become more organized by a global network that will track and regulate all flight movements. It will be like a three-dimensional road system in the sky, but without traffic lights. Ground stations and sky-platforms will provide the mobile radio communications and air traffic control.

Throughout the 21st century the mobile communicator will provide a virtual world for people to use alone or in groups, and while at home or in transit. The transmission radio levels will be less than 10 mW, and often in microwatts. All the human senses will be conveyed by the communications network.

Universal system time will be readily available in all parts of the world enabling the myriad of communication networks to be time synchronized. All mobile radio systems will be based on cellular concepts using spread spectrum techniques. The fixed networks will be packetized, with intelligent packets and network hubs to provide reliable and rapid communications. The OSI model will have long been abandoned. All people will be given a number at birth to use in transactions. The number, in a suitable form, may be embedded into people at birth in order for individuals to automatically respond to a network's request for personal identification. The mobile network ports will be everywhere, and communications will cease to be a problem. There will be no difficulty in locating individuals. It is just as well that we will not be around to experience life in 2200!

REFERENCES

Haskell, B.G. and Steele, R. (1981) Audio and video bit-rate reductions, Proc. Inst. elect. electron. Engrs., **69**, 252–262.

Jakes, W.C. (1974) *Microwaves mobile communications*, John Wiley.

Shannon, C.E. (1948) A mathematical theory of communications, *Bell System Tech. J.*, **27**, July, 379–423 and October, 623–656.

Steele, R. (1985) Towards a high capacity digital cellular mobile radio system, *Proc. Instn. elec. Engrs, Pt F, Communications, Radar and Signal Processing*, **132**, (5), August, 405–415.

Steele, R. (1989) The cellular environment of lightweight handheld portables, *I.E.E.E. Communications Magazine*, July, 20–29.

Steele, R. (1990) Deploying personal communication networks, *I.E.E.E. Communications Magazine*, September, 12–15.

Viterbi, A.J. (1991) Wireless digital communication: a view based on three lessons learned, *I.E.E.E. Communications Magazine*, September, 33–36.

12

Broadcasting after AD2000

C.P. Sandbank, BBC

12.1 INTRODUCTION

An attempt to cover all aspects of broadcasting in a speculation about the next decade is in danger of crossing the border between prediction and science fiction! I have therefore decided to take a more cautious approach by concentrating on a fairly narrow technological field but one which will, nevertheless, bring about a major revolution in the delivery of broadcast vision and sound comparable with that brought by satellites in the present decade.

This paper examines the potential influence of digital broadcasting by drawing attention to current developments and extrapolating from these to their possible application in the 21st century. Digital techniques are coming rather late to broadcasting compared to most other fields of electronics. One reason for this is that semiconductor devices for applications such as a memory or fast fourier transform processing are only just becoming available in a form meeting the demanding technical and price constraints of the broadcast industry. The other reason is the need for compatibility with the existing analogue reception equipment in which the public have made a vast investment which they will not be inclined to write off unless they see a very clear advantage.

Thus the first phase in the application of digital techniques will be the compatible enhancement of the existing analogue broadcast services and this will continue well into the next century as long as there is a substantial consumer base. For example, in Europe, PAL TV services are likely to remain the primary means of domestic TV delivery for the next 25 years, yet in a way, one can already cite established examples of digital technology which has been applied to enhance the existing analogue services. Teletext is a broadcast digital signal which has now been trans-

Communications After AD2000. Edited by D.E.N. Davies, C. Hilsum and A.W. Rudge. Published in 1993 by Chapman & Hall, London, for The Royal Society. ISBN 0 412 49550 3

mitted for many years as a way of augmenting the analogue TV services. NICAM digital stereo sound for TV and the Radio Data System for FM Radio are more recent examples. By the year 2000, improvements to definition and changes in picture aspect ratio from 4:3 to 16:9 will have been established throughout Europe in a compatible manner.

It is always difficult to predict the rate at which a technology feasible in the laboratory will become established in the market place. The remarkable results obtained by techniques like DCT bandwidth reduction and COFDM modulation have demonstrated the feasibility and potential benefits of taking the non-compatible revolutionary step to digital broadcasting and the industry is preparing for its introduction by the year 2000.

The large flat bright high-resolution TV display has eluded the industry which has been actively searching for this product for the last thirty years. Its availability will govern the rate at which HDTV is taken up. But HDTV is only one of the broadcast services providing a market pull for direct digital broadcasting and the introduction scenarios for HDTV and digital TV need not be tightly coupled.

The likely introduction of digital TV at the start of the next century provides yet another opportunity for a worldwide unification of TV transmission standards. It needs to be only loosely related to the existing analogue standards and by next century the world-wide infrastructure of satellite and broadband cable digital distribution will be more apparent and encourage convergence. Unfortunately, it looks as if the TV digital 'revolution' will arrive too early to overcome the commercial and national pressures preventing world-wide unification of standards. Thus the first generation of digital TV is likely to keep its relationship to the existing 50 Hz broadcast standards in Europe etc. and in North America etc., a 60 Hz based digital system is likely. Nevertheless, I propose at the end of the paper a personal view how universality could be achieved, even of it has to wait for the second generation of the digital revolution in broadcasting.

12.2 COMPATIBLE ENHANCEMENTS USING DATV – DIGITALLY ASSISTED TELEVISION

The challenge presented by compatible enhancements is to transmit additional information which can be used by the more complex receiver to display the enhanced picture without impairing the performance of the existing receiver. For the broadcaster, this is an attractive option since his programmes will still be seen by a large audience (albeit without enhancements) during the period in which the new receivers penetrate the market. Further, the existing transmitter system can continue to be used (perhaps with minor modifications). The classic example is the enhancement from monochrome to colour TV where in the PAL system, the colour information is transmitted in a sub-carrier within the luminance band. Although this does cause some loss of luminance resolution around 4.5 MHz spacial frequencies and some 'cross effects' where colour modulates the black and white picture or a stationary luminance pattern can appear as moving colours on the colour display, the impairments are insignificant compared to the enhancement of adding colour to black and white TV.

The task of 'smuggling' say, 1250-line HDTV invisibly through a 625-line transmission is more difficult, but during the coming years one can at least count on very extensive digital processing capability in the domestic receiver. Compatible enhancement using DATV relies on the assumption that one can transmit a substantial digital 'helper' signal with the compatible analogue transmission which will assist the new receiver to recover the 'HDTV' information using bandwidth reduction processes which cause the minimum disturbance to the conventional receiver (Storey, 1986).

The basic concepts of TV bandwidth reduction are firstly, to exploit redundancy in the signal and secondly, to exploit the psychophysical properties of the eye/brain combination to produce the most subjectively acceptable compromises. The tasks which a future digital TV receiver may have to perform can readily be described by reference to Figure 12.1 which shows how the enhanced signal for a 625-line analogue-compatible transmission might be produced.

Starting with say, a 1250-line sequentially scanned picture, this would be analysed to determine the characteristics of the image. For instance, if the picture has areas of fine detail which are not moving, then, assuming the presence of storage in the receiver, the detail can be built up slowly, thus concentrating the use of the transmission spectrum on conveying fine static (or spatial) detail rather than accurate moving (temporal) detail. On the other hand, for an object randomly moving across the screen (e.g., a bird in flight) the eye requires rapidly up-dated temporal information, but is tolerant of a lower spatial resolution in the moving object. Thus in Figure 12.1, two basic types of processing path are shown, one suited to 'moving' images and the other to 'stationary' ones. In addition, the HDTV source is subsampled to make it similar to a 625-line signal. For example, of the 1250-line sequential lines, from the source, every fourth line is transmitted every 50th second so that the original is build up over four fields but would be also recognized as a conventional 625-line interlaced signal with the normal 312-lines per field.

The next part of the encoder simulates both interpolation processes which would be carried out in the receiver to recover something as close as possible to the original from each of the two types of bandwidth-

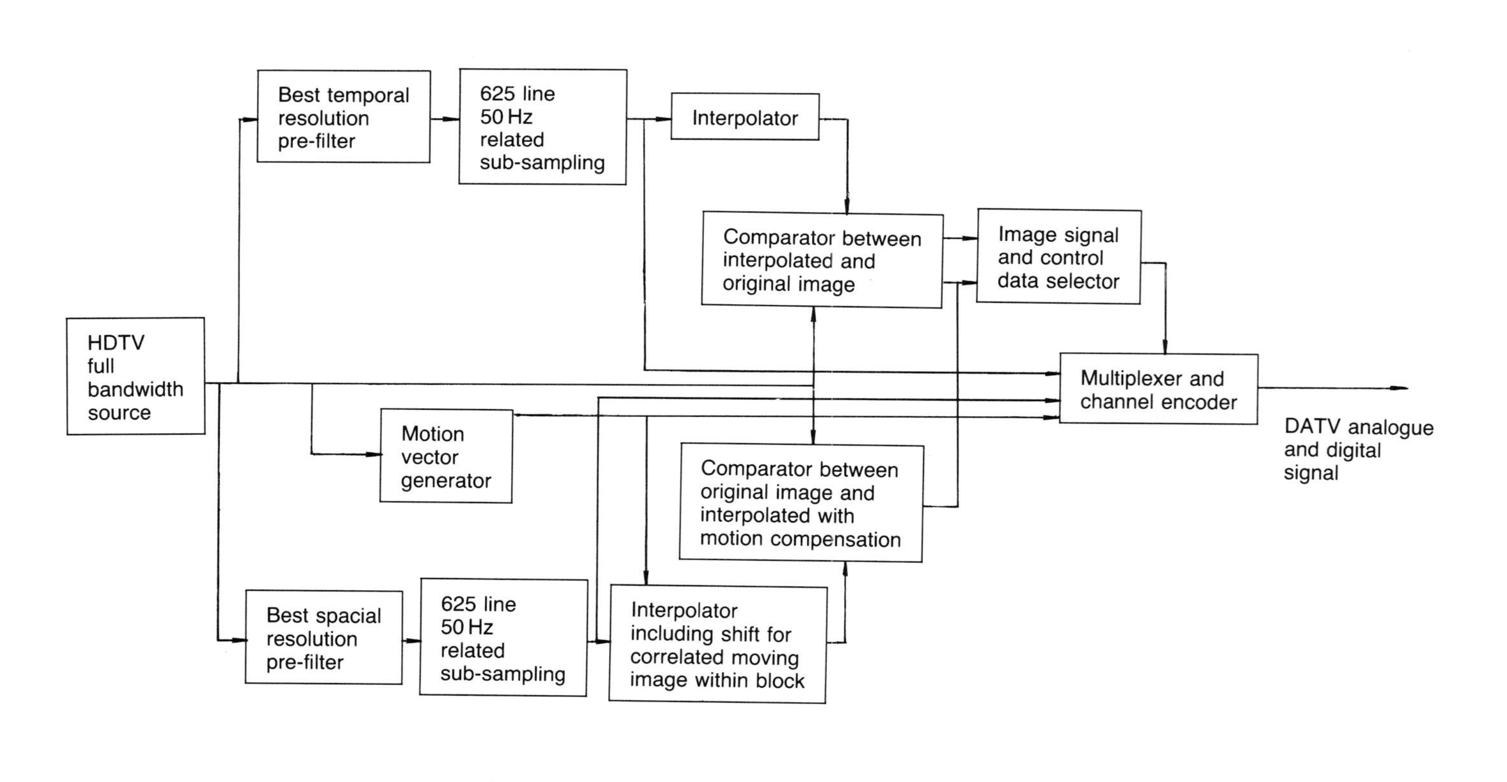

Figure 12.1 DATV encoder for 625-line compatible enhancement.

reduced signal, i.e., using the algorithms designed to suit static or moving detail. The result is compared with the original and the signal giving the best result is chosen for transmission. These decisions are made for each part of the picture in small groups of pixels, and the details of the choice are transmitted to the receiver by the digital assistance signal.

In addition to the broad categories of image, static and moving, there is a third category which has to be dealt with for subjectively acceptable results. Having said that the eye is tolerant of loss of detail in moving objects, this does not apply to 'moving static detail' – an apparent contradiction in terms! If the camera is, for example, showing a hand holding an *object d'art* with fine detail it would not be acceptable if the resolution of fine detail were to diminish as the hand moved the object slowly across the field of view. This is because, unlike the detail in the wings of the bird flapping across the screen, the eye can track the 'moving static detail' of the object if it does not move too quickly. Fortunately, it is possible to deal with this situation without the need for additional bandwidth in the analogue channel provided there is enough capacity in the DATV channel to send the receiver the motion vectors giving the speed and direction of the 'moving static detail'. This enables the object to be treated as if it were stationary with the detail built up over several fields by interpolating the position of the high resolution images in the intermediate fields from a knowledge of the speed and direction of the moving detail. This can be done with an accuracy of better than 0.1 of a pixel using phase correlation to generate the motion vectors (Thomas, 1987). Thus the sophisticated receiver would take the analogue signal and convert it to a digital signal suitable for processing. Then, using the DATV signal, it would process the signal in one of the three ways judged by the encoder to be likely to give the closest approximation to the HDTV original.

This is essentially the process which will be introduced later in this decade when HDMAC services are added to the D2MAC satellite transmission which have been started by several European DBS operators. An additional complication is that the new services will use the HDTV format of 16:9 (Figure 12.2) compared to the conventional 4:3 format. Fortunately, the D2MAC 625-line DBS transmission standard makes provision for the 16:9 format and the only additional signal which must be sent to the conventional receiver is which part of the 16:9 frame the broadcaster wants the 4:3 viewer to see have displayed on his screen. (In the absence of this signal the central portion will be displayed).

A similar process of enhancement is also possible for the terrestrial PAL transmissions but with some additional difficulties. Firstly, because the PAL signal is composite (i.e., the colour sub-carrier is within the luminance band) it is necessary to use up-to-date, more efficient means of

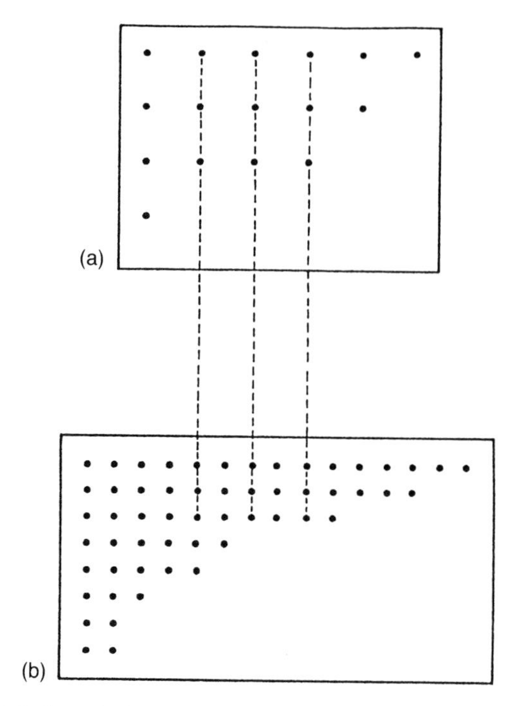

Figure 12.2 Relationship between Rec 601 and HDTV digital sampling structure.

separating the luminance from the chrominance than were possible in the days before digital receivers. Furthermore, this must be done without confusing the conventional PAL decoders. Secondly, the terrestrial TV channels have a narrower bandwidth than the DBS channels so that there is less opportunity for resolution enhancement. Thirdly, since the existing PAL system did, of course, not anticipate wide screen display this presents a special problem.

The problem of achieving compatibility between the 16:9 and 4:3 aspect ratios is likely to be one of the most difficult during the next ten years. It is rather more of an artistic than a technical problem. In the case of PAL, there are two main technical options. One can add 'sidepanel' information somewhere in the signal and 'stitch' these on the wide screen receiver. Alternatively, one can transmit all signals so that they appear in 16:9 'letter-box' format like 'cinema-scope' films on the 4:3 screen so that they can fill the 16:9 screen on the enhanced receiver.

The artistic problem is that for the next twenty years TV programmes will have to be produced with two formats in mind. If the producer uses a Hollywood technique of 'shoot and protect' (i.e., the camera man makes sure there is nothing of importance in the sidepanels) then the whole artistic advantage of the new format is lost. If he ignores the possibility of 4:3 transmissions, then there may be something essential to the action in the sidepanel which the conventional viewers do not see. Whilst to some extent this can be dealt with by 'pan-scanning' in pre-production before transmission, this is not really possible for live programmes such as sport, news and current affairs etc.

That is why in the 'PALplus' enhanced terrestrial project the letter-box solution has been adopted, even though this effectively reduces the resolution of the picture sent to the conventional viewer and therefore does not strictly meet the requirement not to impair the existing service. Figure 12.3 shows how the various inputs to a compatible PALplus service may be handled as we move into the next century. By that time, an increasing proportion of the programme material will be in 16:9 component form at 576 active lines and 720 pixels per line (a) and then in the HDTV format of 1152 lines 1920 pixels per line (b). Although a compatible enhanced system like PALplus is unlikely to do justice to HDTV sources, by around 2010 broadcasters will probably only produce in HDTV in order to satisfy the demands of all the services from HDTV to enhanced and conventional PAL with one single format having enough 'headroom' for all programme outlets.

However, there will still be a large amount of 4:3 conventional PAL material (c) e.g., from 'un-converted studios' (the normal BBC studio refurbishment cycle is ten to fifteen years) archival programmes and news sources using conventional cameras etc. This will first have to be 'clean decoded' (d) to remove cross effects and 'cropped' to the 'letter-box' format of 16:9 which would illuminate 432 of the 576 active lines on a 4:3 format (e). These 432-lines would then have to be expanded to 576 to fill a 16:9 format (f) before processing by the PALplus encoder (g). Alternatively, if the producer did not want to lose the top and bottom of the original image, the sides of the 16:9 frame can be left black thus displaying all 576-lines (h).

Either of these alternatives would be processed by the PALplus encoder which would convert the image to 432 standard PAL compatible lines plus 144 lines containing helper signals (i). These make use of the black bands in the letter-box image to send information which can be recognized by the 16:9 receiver (j) and used to recover as much as possible of the original 576 or 1152-line image but are below black level and will not show on the conventional receiver (k).

There is still a question mark about the extent to which British audiences will accept a letter-box presentation on their 4 × 3 receiver. Unlike their continental counterparts, British broadcasters tend to 'pan-scan' cinemascope films as a matter of course to fill the screen. There may therefore have to be a transition scenario where UK audiences may be

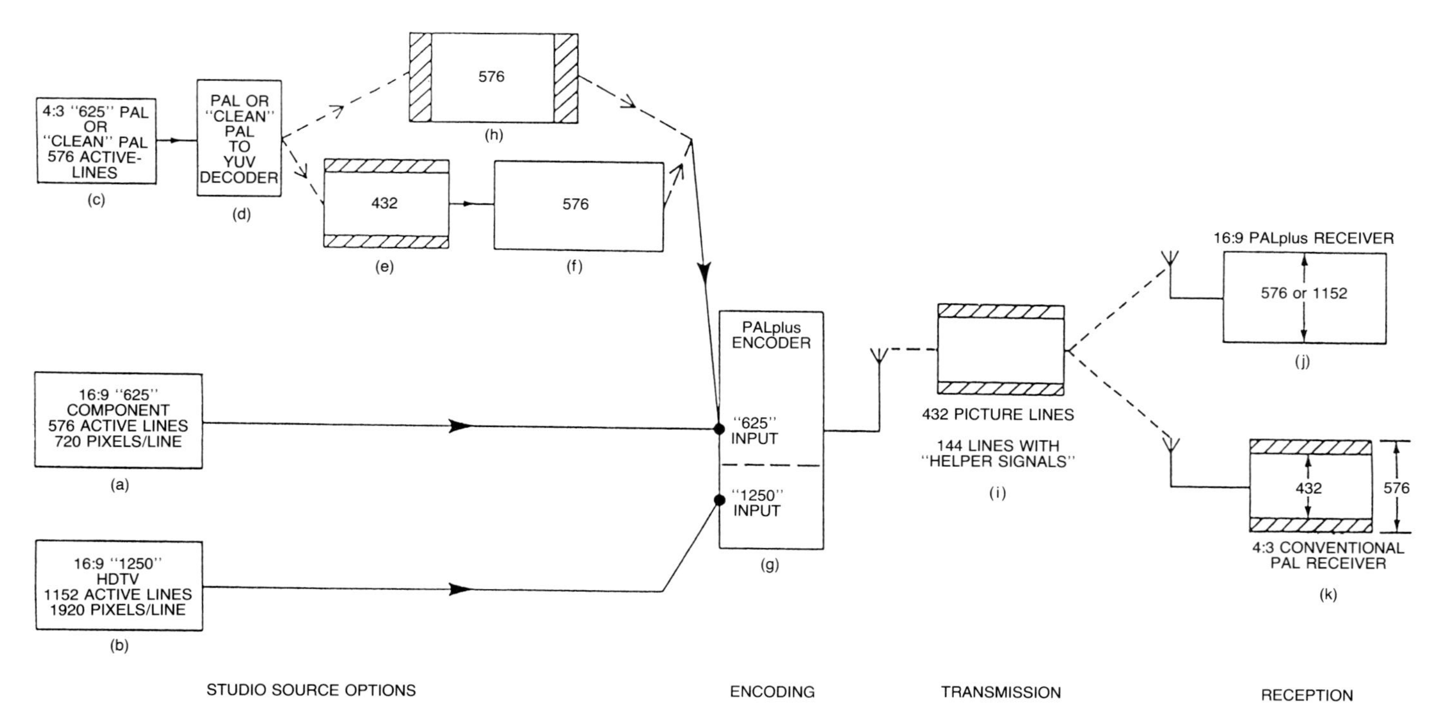

Figure 12.3 Options for PALplus services.

gently led to 16:9 through a 14:9 intermediate stage (without 'helper' signals) before a regular PALplus services could be widely introduced.

12.3 DIGITAL SOUND BROADCASTING

Before returning to the subject of TV, it is appropriate to deal with digital sound broadcasting as this will hopefully be established by the start of the next century and some of the concepts are also applicable to digital TV broadcasting.

It has been a source of annoyance to radio broadcasting executives that in the UK, TV viewers should be able to enjoy digital stereo sound of 'CD' quality, by courtesy of the NICAM system (ETSI, 1992), while radio has to make do with the over thirty-year-old FM system as the Hi-Fi member of the family. There are two reasons why digital sound for TV came before radio. Firstly, TV receiving antennas are fixed and not subject to significant variable multi-path. Secondly, compared to the video signal, audio requires a modest bandwidth and space could be found to squeeze it into the existing TV spectrum with the modest bandwidth reduction of NICAM which requires 728 Kbit s^{-1} for the stereo sound channel.

Although the FM network was envisaged for fixed antenna reception (it pre-dates the 'transistor portable') mobile reception at home and particularly in cars has now become an essential part of the service. It is essential therefore to solve the variable multipath problem before contemplating digital radio broadcasting. Further, it is necessary to devise ways of bandwidth compression to increase the chances of finding spectrum dedicated to audio broadcasting. The reduction of the required data rate for high quality sound also contributes to the feasibility of efficient multipath-insensitive coding.

Using the terminology symbol period (rather than bit period because with multi-level coding i.e., 4-phase differential phase shift keying (DPSK) there are two bits coveyed per symbol period) if a broadcast digital stream is not to be affected by multipath the reflected signal must arrive within a fraction of a symbol period. For example, in the case of NICAM 728 using 4-phase DPSK the signal period is 2.7 μs, or 800 m in distance, and to avoid interference the added distance travelled by the multipath reflected signal should be less than 25% of the symbol length, say 200 m. In practice, the signal would have to cope with path differences of about 100 times this distance.

The solution adopted by the European collaborative project for digital audio broadcasting (DAB) (Pommier *et al.*, 1990) is to reduce as much as possible the bit rate for high quality sound as a contribution to lengthening the symbol period and then use an elegant spread spectrum technique

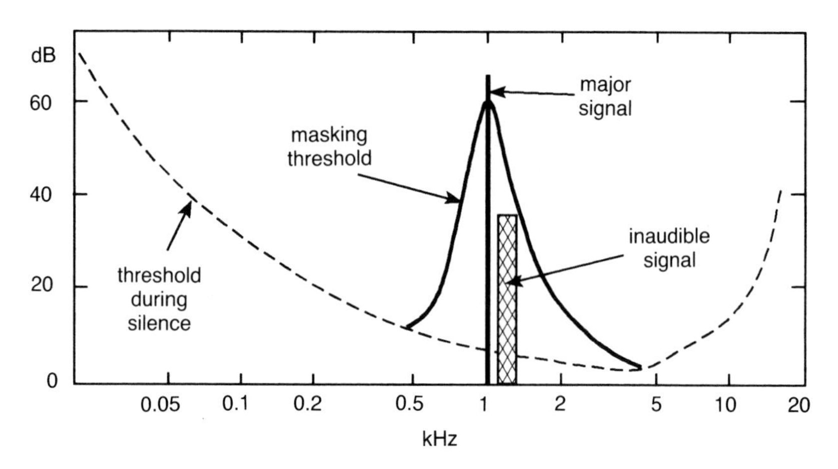

Figure 12.4 MASCAM principle.

to multiplex the signal onto many separate carriers modulated by digital signals with symbol periods of over 1 ms.

The techniques of bandwidth reduction relies on the psychophysical phenomenon that whilst the ear appreciates the high dynamic range offered by digital sound, loud sounds can mask those which are quieter. Thus the audio spectrum is divided into sub-bands which correspond to the width of the masking threshold shown in Figure 12.4. each of which is separately quantized using the appropriate scale factors for a short period (similar to NICAM). However, a much coarser quantizing law is used to reduce the bit rate and this is adequate for low levels of sound on their own in the sub-band. If there is a loud signal in the sub-band, then there will be quantizing noise but this is not audible because it appears only in the sub-band and will be masked by the main signal. Any low level signals in the presence of a loud signal will be discarded by the coarse quantizing but this is no loss since they would have been masked in the ear by the loud sound in the sub-band. This technique named MASCAM (masking pattern sub-band coding and multiplexing) (Thiele *et al.*, 1988) has given excellent results at 128 Kbit s^{-1} for 20 KHz sound which represents a 6:1 bit-rate reduction from linear coding.

The spread spectrum technique proposed for DAB is COFDM (coded orthogonal frequency division multiplexing) and is particularly spectrum efficient because several programmes are transmitted in the multiplex and there is the possibility of a single frequency National Network. The basic principle of COFDM is to use a fast Fourier transform (FFT) to perform the tasks of spreading the relatively high data rate of the digital audio signal into a large number of carriers with a much longer symbol period.

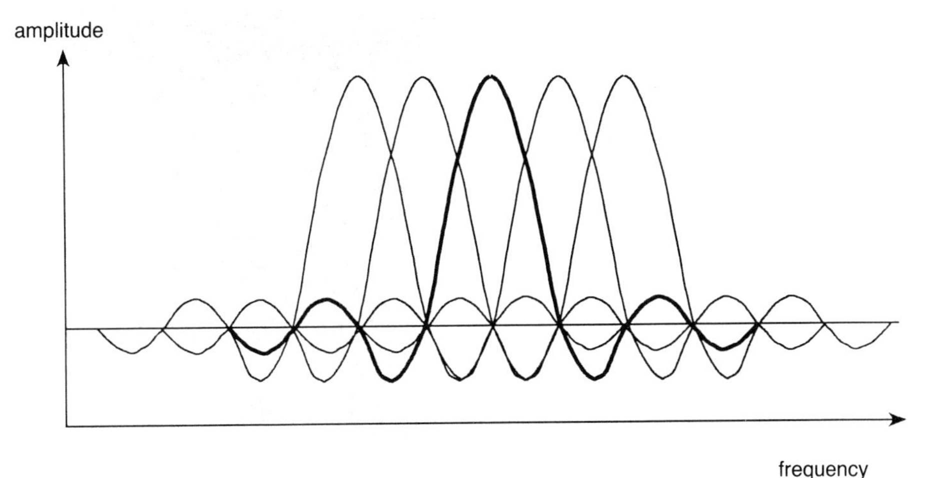

Figure 12.5 COFDM interleaved carriers.

Whilst there are many ways in which the signal could be distributed among many carriers the advantage of using FFT is that it produces a series of carriers which, although overlapping as shown in Figure 12.5 have the property of mathematical orthogonality. This means that when the inverse FFT algorithm is used in the receiver to decode the spread spectrum signal it will perform a correlation with the transmitted spectrum in which the fact that the carriers overlap will not contribute to crosstalk.

The conditions of orthogonality will only apply if all the states of the carriers defined by the discrete Fourier algorithm in the encoder are detected by the receiver for the complete symbol period. However, since the system is designed to cope with delays due to multipath and the delay may vary from carrier to carrier, it is necessary to add a guard interval t_g to the active symbol period such that the total symbol period

$$T_s = t_s + t_g$$

where t_g is about 25% of t_s.

This means that although the states of the carriers (i.e., their phase in the case of PSK) is maintained for a period T_s, the FFT process in the encoder at the transmitter and in the receiver are both carried out over a period t_s during which phase continuity and hence orthogonality will be maintained.

Typical values envisaged for a service to be implemented later this decade are T_s = 1250 μsec and 3000 carriers for a 12 stereo programme service.

Figure 12.6 is a photograph of the actual COFDM spectrum during a

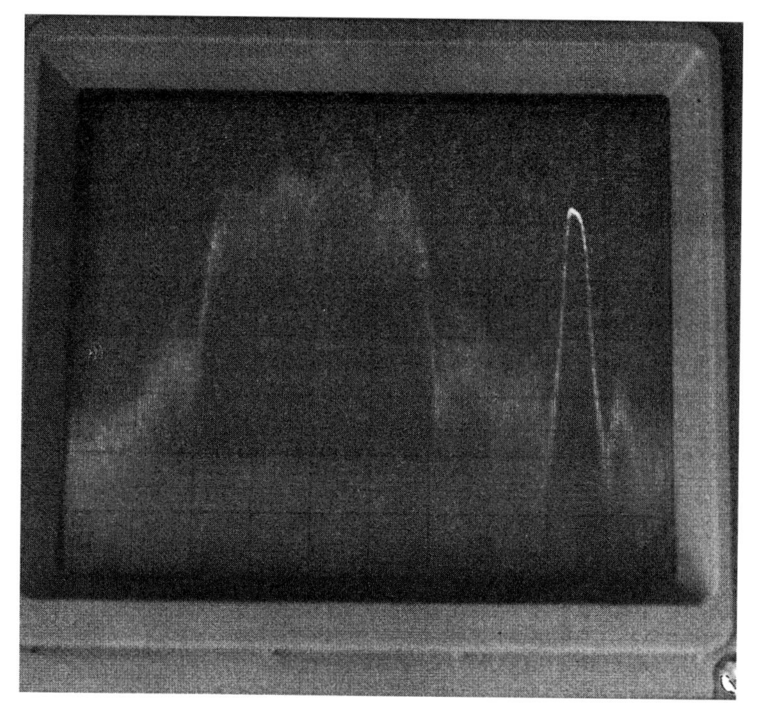

Figure 12.6 Spectrum of COFDM signal.

mobile reception test of an experimental transmission comparing DAB with FM performance. The FM carrier can be seen at the right of the spectrum. In a mobile situation in the presence of multipath, carriers close in frequency will be affected in the same way for short time intervals but information sent on carriers further away in frequency or later will be affected independently. This is illustrated in Figure 12.7 where the small squares represent T_s and adjacent carriers and the large squares enclose an area which would be affected independently from the area enclosed by another large square. The individual elements linked by the diagonal lines represent consecutive bits of the encoded audio signal. These are likely to be in areas affected differently and this time and frequency diversity, combined with powerful error protection strategies for the transmitted signal will make future DAB services extremely rugged and free from apparent interference. It should be noted that providing a reflected signal arrives with a delay of less than t_g it will make a constructive contribution to the signal. In fact it is not until there is a delay of $2t_g$ that the interference is destructive although there is little improvement in signal-to-noise ratio from signals delayed between t_g and $2t_g$. Thus signals from

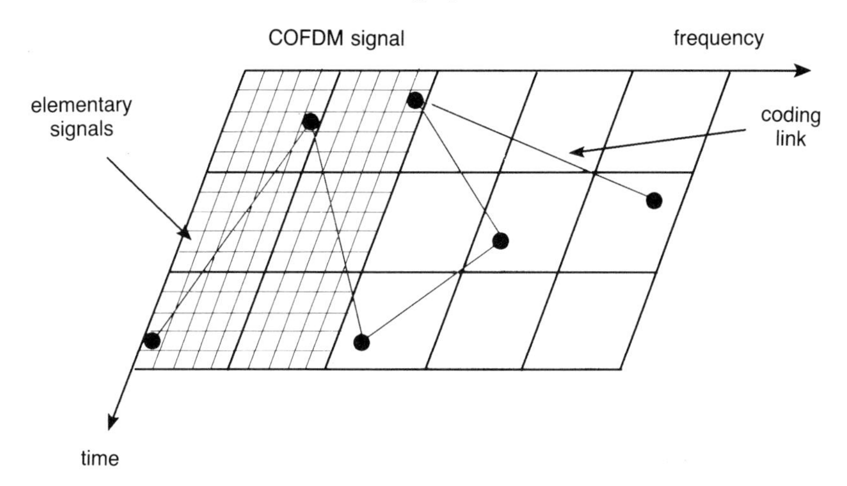

Figure 12.7 Relationship of individual carriers to propagation conditions.

distant transmitters broadcasting the same programme spread over the same frequency spectrum can contribute to received signal strength rather than causing 'co-channel' interference which must be avoided in planning the FM service. In the case of the example quoted earlier where t_g = 250 µsec signals from transmitters up to 70 km away would contribute constructively. This means that provided the transmitters are synchronized, a single frequency network can be established which might provide 12 stereo programme services in a total spectrum of 3 MHz (say 3.4 with guard bands). Currently to provide one stereo pair on a national basis in the UK, 2.2 MHz are required to avoid co-channel and adjacent interference.

The high quality audio reception which DAB offers coupled with its efficient use of the radio spectrum make it very likely to become the dominant radio delivery means in the next century. Eventually portable receivers are likely to be configured so that they can be used with terrestrial and satellite transmissions.

12.4 DIGITAL TELEVISION BROADCASTING

Although the same principles as used for DAB should eventually be applicable to the direct broadcasting of digital television, this is in a much earlier stage of development and the problems are more severe. This can be quantified by reference to Figure 12.2. Early next century, a digital TV delivery system would be expected to be capable of delivering HDTV. Using the 1920 pixels per line format of Figure 12.2 with twice the number of lines in the '625'-line representation of CCIR Rec 601 (Sandbank, 1990, pp. 8–16), one can derive the raw bit rate for HDTV. Rec 601 specifies 8 bits per sample at a sampling frequency of 13.5 MHz per luminance and 6.75 MHz for each of the two chrominance signals thus requiring a data rate of $216\,\mathrm{Mbit\,s^{-1}}$. The HDTV raw data rate is therefore

$$\frac{1920}{720} \times 2 \times 216\,\mathrm{Mbit\,s^{-1}} = 1.12\,\mathrm{Gbit\,s^{-1}}$$

In terms of the symbol lengths and spectra discussed in the previous section even the $216\,\mathrm{Mbit\,s^{-1}}$ of Rec 601 would present a formidable problem, but the requirements for HDTV emphasize the point. There is no doubt that the practical realization of digital TV as a consumer product will require elaborate signal processing by techniques still to be invented. However, the remarkable and consistent results which have been achieved by recent digital TV band-width reduction techniques makes it worthwhile drawing attention to this as the enabling terminology which will bring digital TV delivery to the market place early next century. Algorithms based on the DCT (discrete cosine transform) have produced the most spectacular results. Starting with $216\,\mathrm{Mbit\,s^{-1}}$ signals, bit-rate reductions of 200:1 have been achieved with picture quality comparable with that achieved by the average domestic video cassette recorder. Bandwidth reduction of 50:1 gives results which have some impairments but are subjectively very close to the original. At 25:1, it is extremely difficult for even 'expert' viewers to spot the impairments on critical pictures and one can consider the system 'transparent'.

Table 12.1 illustrates how the DCT contributes to bit–rate reduction by taking the typical 8×8 block used in bit-rate reduction and examining a simple pattern of alternate black and white pixels. The DCT has a distribution of coefficients such that the majority are at or near zero (for a mathematical treatment see Thiele *et al.* (1988, pp. 610–617). In the example shown, by setting all coefficients below 64 to zero, one can see from the lower output matrix that this is still fairly close to the original. This is visually presented by the computer simulation in Figure 12.8 where each pixel has been 'enlarged' on the screen for clarity. Figure 12.9 shows how this process appears on a typical picture rather than the simple check pattern and in Figure 12.10, twice the difference between the original signal and the inverse DCT is shown to indicate the error introduced by the DCT as fewer significant coefficients are used.

The DCT is most effective in blocks containing high spatial frequency components. Figure 12.11 is a different check pattern having lower spatial frequencies than Figure 12.8 and the inverse transform is not quite so effective. It may seem that differential PCM (DPCM) might be better.

Table 12.1 DCT example in 8 by 8 blocks

Input matrix:								
	255	0	255	0	255	0	255	0
	0	255	0	255	0	255	0	255
	255	0	255	0	255	0	255	0
	0	255	0	255	0	255	0	255
	255	0	255	0	255	0	255	0
	0	255	0	255	0	255	0	255
	255	0	255	0	255	0	255	0
	0	255	0	255	0	255	0	255
Transform matrix:								
	1020	0	0	0	0	0	0	0
	0	33	0	39	0	58	0	167
	0	0	0	0	0	0	0	0
	0	39	0	46	0	69	0	196
	0	0	0	0	0	0	0	0
	0	58	0	69	0	103	0	294
	0	0	0	0	0	0	0	0
	0	167	0	196	0	294	0	837
Output matrix:								
	255	0	255	0	255	0	255	0
	0	255	0	255	0	255	0	255
	255	0	255	0	255	0	255	0
	0	255	0	255	0	255	0	255
	255	0	255	0	255	0	255	0
	0	255	0	255	0	255	0	255
	255	0	255	0	255	0	255	0
	0	255	0	255	0	255	0	255
Now set all dct terms less than 64 to zero								
New transform matrix:								
	1020	0	0	0	0	0	0	0
	0	0	0	0	0	0	0	167
	0	0	0	0	0	0	0	0
	0	0	0	0	0	69	0	196
	0	0	0	0	0	0	0	0
	0	0	0	69	0	103	0	294
	0	0	0	0	0	0	0	0
	0	167	0	196	0	294	0	837
Output matrix:								
	207	−2	257	−6	261	−2	257	48
	−2	276	9	250	5	246	−21	257
	257	9	249	−10	265	6	246	−2
	−6	250	−10	248	7	265	5	261
	261	5	265	7	248	−10	250	−6
	−2	246	6	265	−10	249	9	257
	257	−21	246	5	250	9	276	−2
	48	257	−2	261	−6	257	−2	207

Figure 12.8 Input and output check patterns.

Figure 12.9 Pictures coded by retaining only the amplitude coefficients in an 8×8 DCT. The number of coefficients per block is as follows:

64	56	48
32	24	16
8	4	64 (original picture).

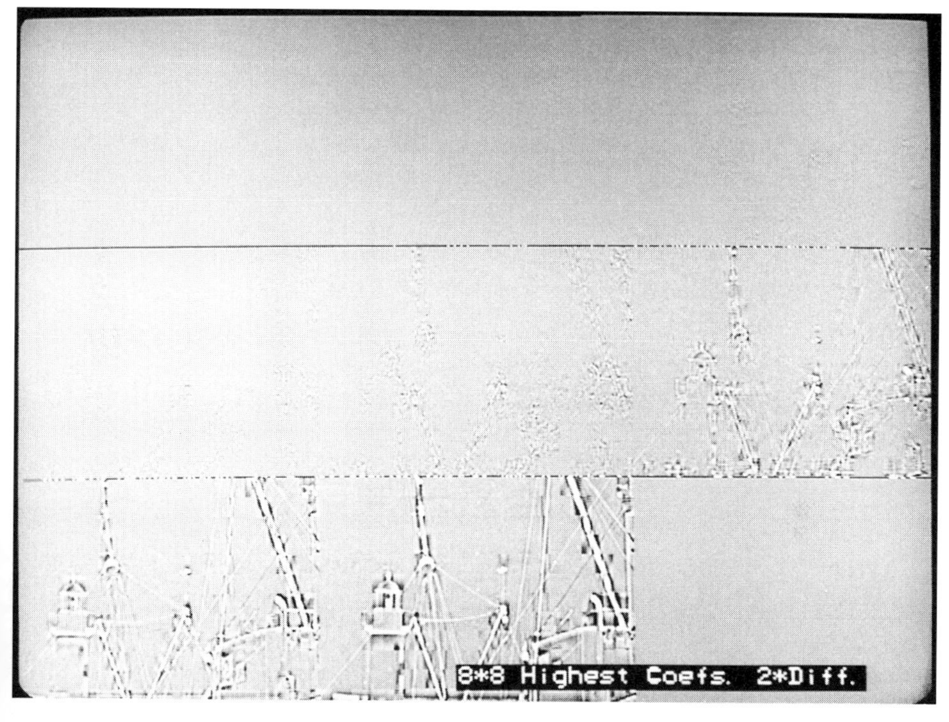

Figure 12.10 Picture showing the difference between the coded pictures and original for the sequence of pictures in Fig. 12.9.

Figure 12.11 Coarse check pattern.

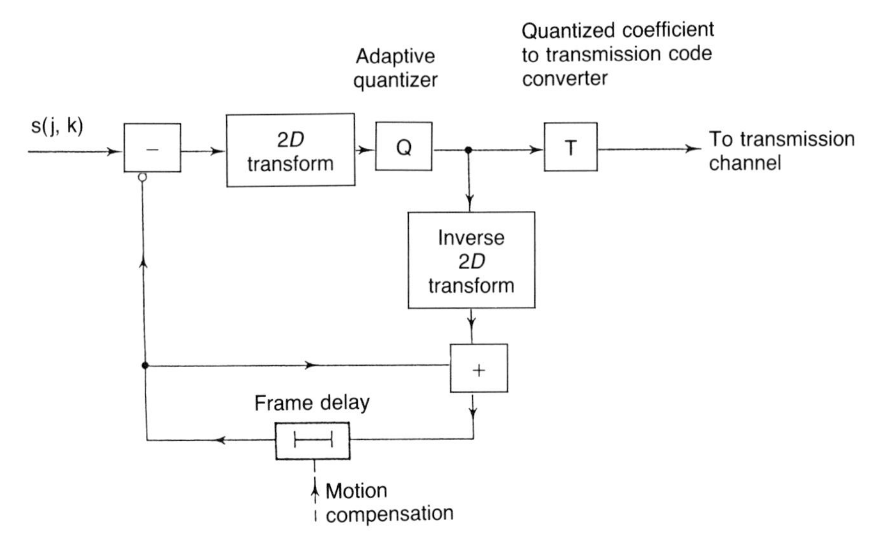

Figure 12.12 Block diagram showing principles of hybrid predictive-transform coding.

DPCM is the process where only the difference between the actual and predicted values based on previous samples in space or time is used. In practice however, for spatial bandwidth reduction the DCT always seems to win.

When it comes to the temporal domain, DCPM gives the best results and the sequence for this 'hybrid' approach can be seen from Figure 12.12. The samples in the two dimensional blocks are transformed into DCT coefficients and only the most significant values from the adaptive quantizer Q are then used to obtain the inverse transform for comparison with the predicted image from the previous frame so that the DPCM signal can be derived. Where there is strong correlation between the image in the blocks from consecutive frames due to the 'moving static detail' discussed in section 12.2, then a motion compensated DPCM term is used giving a further saving. In the same way as discussed with reference to Figure 12.1, the coder would compare various options for coding the block with the original to select the best compromise between least error and maximum bandwidth reduction. Non-linear quantization is used to encode the more significant high values which occur less frequently with greater accuracy (i.e., more bits per sample) than the more common low values. Finally variable length coding is used at T to obtain further savings in bandwidth. A buffer store is used to cope with the variation in the data rate and a feed-back loop to Q can be used to cope with buffer saturation.

With the possibility of reducing the bit-rate of HDTV signals to less than 20 Mbit s^{-1} and the cost of VLSI coming down to make parallel processing of the high speed transform viable, there seems little doubt that we will see digitally delivered HDTV early in the next century. Other applications of digital TV offering better utilization of the spectrum or more rugged reception for portable receivers may come earlier.

12.5 THE DIGITAL NETWORK

When engaged in a similar crystal gazing exercise in 1979 (Sandbank, 1980*a*) I wrote

> I am confident that during the 21st century we should see the common carriers providing a very wideband switched communications network . . . I am less confident about the timing . . . it may be well into the 21st century before we can count on the extensive penetration of a network suitable for video broadcasting and communication.

Twelve years later and much nearer the 21st century, I am no more certain about the timing for the penetration of a broadband integrated subscriber digital network to an extent comparable with the domestic availability of telephone or UHF TV today. The problems of optical fibre transmission which were then challenging (Sandbank, 1980*b*) are now solved. The problems of switching have consequently risen to the top and present some formidable obstacles. Nowhere is it more evident than in the case of a TV studio network which aims to cope with a future digital TV infrastructure. It can be regarded as a customer premises network (CPN) having typically some 200 terminals each terminal producing or accepting signals of over 1 Gbit s^{-1} in the case of an HDTV operation (Marsden *et al.*, 1990).

Whilst bandwidth reduction can and will be used, unless transparent encoding and decoding can be carried out many times in sequence (see section 12.6) the process of studio post processing such as colour separation overlay (Sandbank, 1990, pp. 539–556) will require access to the full bandwidth TV signal. The magnitude of the problem can be illustrated by the solution under consideration for a future TV studio centre. This consists of an optical fibre switched network using both wavelength and time division multiplexing.

A WTDM broadband CPN envisaged for a conventional digital TV network could have a number of local routeing centres (LRCs), each serving as interfaces between the main core of the CPN and the customer's terminal equipment. Figure 12.13 shows a system with sixteen LRCs each serving sixteen 155 Mbit s^{-1} sources and sixteen destinations. This gives a total of 256 sources and 256 destinations.

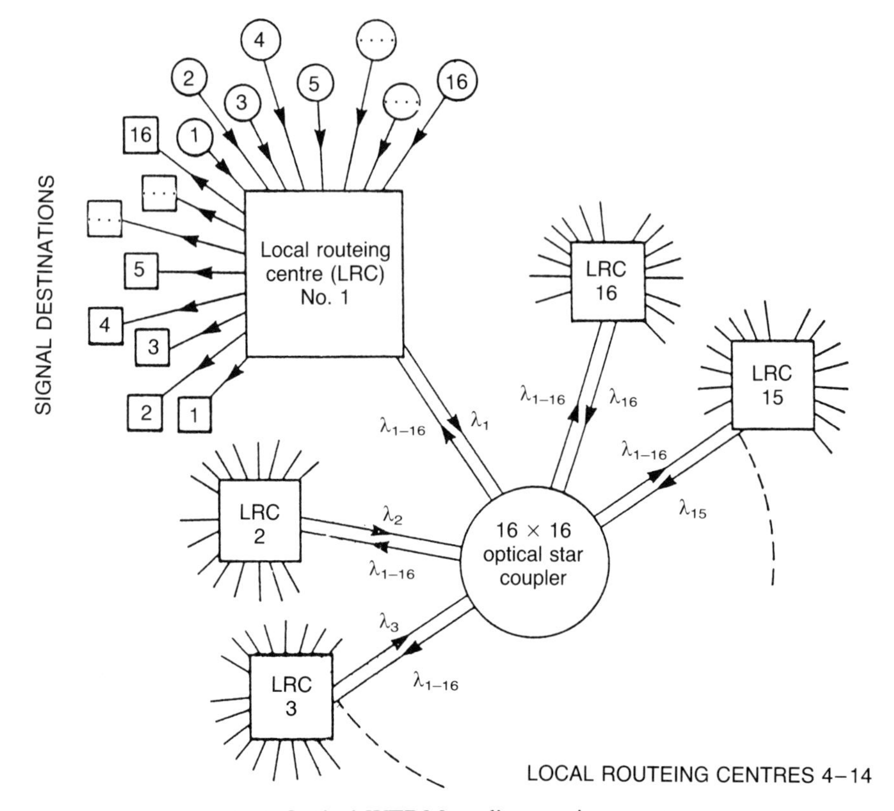

Figure 12.13 Optical WTDM studio routeing arrangement.

At each LRC, the digitized signals time-multiplexed electronically to form a serial bitstream at up to 2.5 Gbit s^{-1}, following CCITT Recommendations G 707–9 for the synchronous digital hierarchy (SDH). The 2.5 Gbit s^{-1} serial signal modulates a distributed feedback (DFB) laser to generate an optical signal at a controlled wavelength. A different wavelength is used at each LRC, the wavelengths being equally spaced 4 nm apart between 1500 nm and 1560 nm.

The outputs of all the LRCs are transported using single-mode optical fibres to a centrally-located optical star coupler where they are combined to produce several identical high-density wavelength multiplexed optical signals. These optical signals are then distributed via more single-mode fibre to each of the LRCs where they are demultiplexed optically using grating-based demultiplexers and then demultiplexed electrically to provide access at each of the destinations to any of the sources in the system.

The CPN would be provided with a control system which will allow the selection of any source at any destination without centralized switching. For each specific application user control software will be needed to translate the customer's control procedures into the commands required to control the CPN.

One or more of the LRCs could form an IBCN interface to exchange signals with the broadband public network following CCITT recommendations. This interface will include the exchange of signalling between IBCN signalling channels and the broadband CPN control software.

Several different configurations of the broadband CPN would be possible to suit different applications. Networks could be connected in series or in parallel, either to provide more sources and destinations or to interconnect networks in different areas.

Once outside the BCPN, it will depend on the penetration of the broadband ISDN (BISDN) to what extent different BCPNs can be connected with the ease of setting up a telephone conversation today. It may be that the establishment of effective bandwidth reduction process at consumer product prices would hasten the use of BISDN to deliver broadcast services to the home on digital bearers. The widespread use of fax may create a market for the photographic equivalent followed by the desire to send video 'messages' with the ease with which we send a fax today. This may lead to the gradual establishment of the continuum from telephony to full TV quality transmission on the domestic network and this would certainly be facilitated by a common digital multiplex for all the applications. However, it seems unlikely that this could be the norm before the end of the first quarter of the next century.

12.6 A UNIVERSAL APPROACH TO DIGITAL TELEVISION

12.6.1 Background

When standards for HDTV were first proposed, broadcast engineers hoped that the move to HDTV might provide the opportunity for a break with the 50/60 duality which was bequeathed by the Victorian power engineers. Sadly, for a number of reasons, some of them soundly based on engineering considerations, others political, the duality is likely to be preserved for the first generation of HDTV activities which are envisaged for the end of this decade.

A move to direct digital television transmission offers us one more chance to try to achieve a universal approach for programme making, interchange, and delivery to the public. The main obstacle to be overcome is to achieve independence from the rigid concept of field frequency and the operation of the system, and a possible approach is discussed below.

In the digital bandwidth reduction systems discussed earlier, it is common practice to send information according to algorithms which relate the frequency at which picture detail is sent to the nature of the image. This ranges from a few pictures per second for 'stationary' (or motion compensated) detail which is sent at high resolution, to fast moving detail which is sent frequently at lower resolution.

Thus, the first step in decoupling from the fixed field rate has already been taken. Whilst it is true that synchronous systems are effective and the various periodicities tend to be harmonically related to the field frequency, the choice of 'repetition' rates in a digital bandwidth reduced system is based primarily on what is needed to allow the best subjective reconstruction in the receiver.

The display device of the future may require information update in a form very different from the CRT. We have already seen liquid crystal projection displays giving bright high resolution pictures using a technology which can cope with flicker elimination at the display element level, therefore invalidating the arguments about the need to update the display at a frequency related to the persistence of vision of the eye. Updating would be on the basis of the algorithm for best subjective portrayal of fine detail and movement, which does of course depend upon the characteristics of the eye, but is much more sophisticated than a consideration of the response of the edge of the retina to flicker in bright light!

It is already becoming the practice to use or ignore certain elements of the emission depending on the cost of the receiver. Thus a future universal emission standard could also be configured in such a way that it could support a wide range of receivers from the low-cost portable low-definition receiver to the sophisticated high-definition receiver as well as other applications such as fax and teleconferencing.

12.6.2 Possible system approach

In a *reductio ad absurdum* scenario one could consider a system in which the rigid discipline of a periodic scanning structure is dispensed with by sending new information from each pixel only on occasions when a temporal change made this worthwhile. A header with the information would make sure that the result was displayed in the correct way and position. In practice such a system would probably require more bandwidth to transmit the headers than that which was saved by any bandwidth reduction algorithm! Thus some sort of regular spatial quantization such as scanning lines which define the location of large groups of pixels would be helpful. Regular temporal sampling is also helpful to enable change to be measured readily for subsequent processing. However, if the eventual signal is to contain only the essential information sent in a way not constrained by quantized spatial or temporal sampling then the original sampling must be carried out with plenty of headroom i.e. several times the highest resolution required of the system.

Whilst such an approach at first sight would seem to require astronomical raw data rates for HDTV before the first stages of processing, we might reduce the problem to manageable proportions by using the parallel processing methods similar to those believed to be the basis of human visual perception (Spillmann and Werner, 1990).

With this in mind, Figure 12.14 illustrates a possible universal system approach. The main difference between this and our present concept for digital television is that the sensor module is integrated with the first stage of signal processing and bandwidth reduction. This enables one to be 'very extravagant' with headroom at the first stage of source coding. Thus, the information fed from the sensor to what I have called the 'sensor module' could have spatial and temporal resolution which might require an equivalent of several hundred Gbit per second in a serial stream without making particularly onerous demands on processing technology because the information would be fed in a large number of parallel streams from one part of the integrated circuit to another.

From here on processing can follow some of the established techniques of sensing and transmitting difference signals, but rather than tying this to differences between 'frames', it would be adapted to the nature of the information in the blocks of the picture, as illustrated in the lower part of the diagram. Thus, the information from a particular block would be sent infrequently at high resolution or combined into single resolution elements sent much more frequently, but the decisions for this 'conditional replenishment' would be made on the basis of the nature of the image, the bandwidth available at the next stage of processing and the purpose for which the information will be used. The choice of the appropriate algorithm would be determined by these three considerations and would be fed into the sensor module to enable the appropriate means of processing to be selected. Coming from the sensor module would be the 'bandwidth reduced' picture information, coupled with control information stating which methods of processing had been chosen for the blocks, plus motion vector information, etc.

With this information the signal can be further processed and bandwidth reduced, say, for recording or transmission, or it can be 'expanded' for post-production or ultimate display. Of course the expansion process cannot replace information which has been irrevocably removed but it can, using the combination of picture information and control informa-

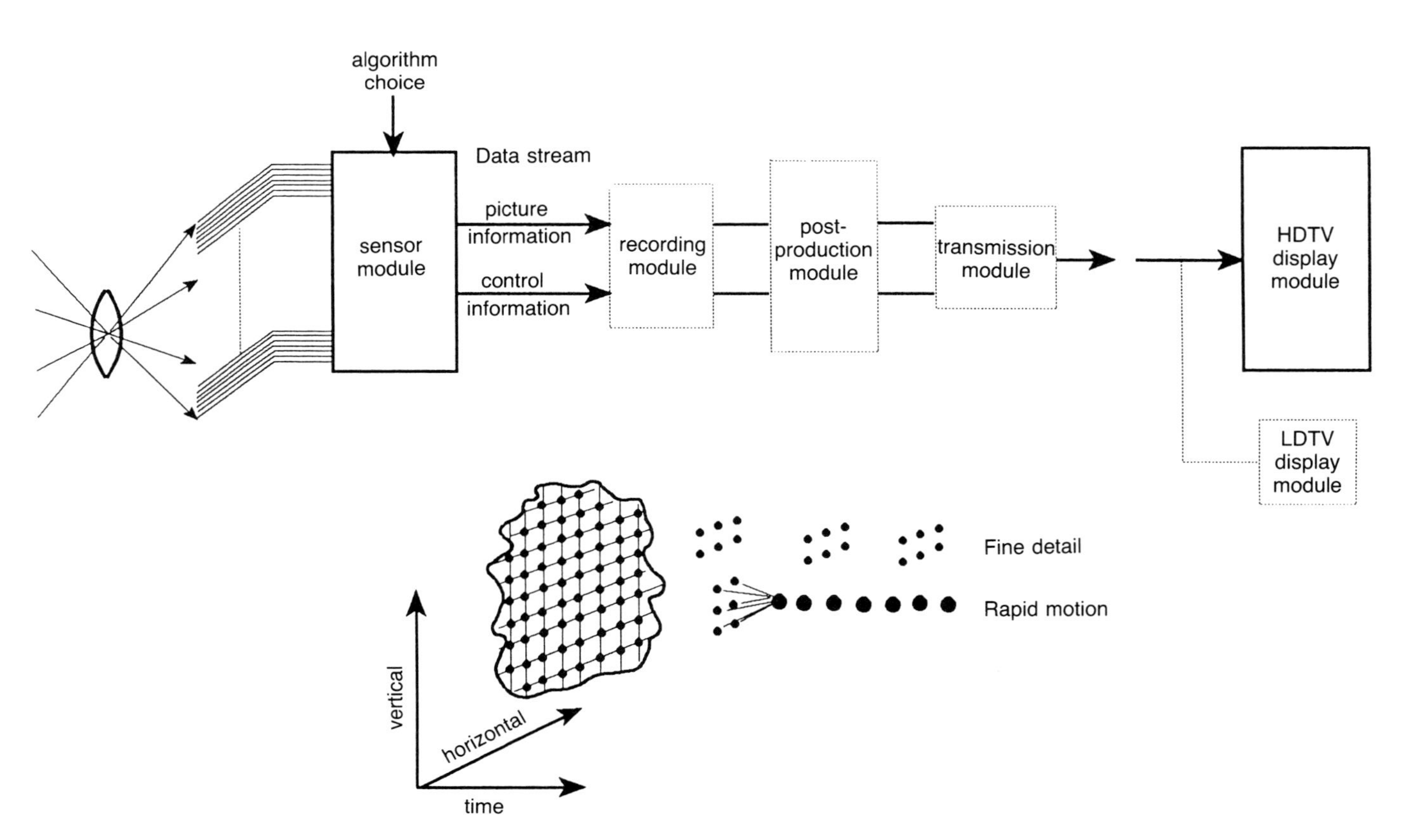

Figure 12.14 A universal approach to digital encoding.

tion, subjectively 'expand' the picture detail by taking advantage of the fact that, for example, in post-production and in display the bandwidth limitations of recording and transmission do not apply.

Extreme examples of differences in algorithm choice which might be fed into the sensor module are, on the one hand, the algorithm appropriate for facsimile transmission, i.e., static but very high spatial resolution, and on the other hand for recording of sports events, some of which may be used for high definition slow-motion presentation which need high temporal resolution.

The hierarchical approach at the broadcast receiver would mean that the signal would be arranged in such a way that the top of the range high-definition display module would use most of the information in the digital transmission, whilst the low-cost 'low-definition television receiver' would have a display module extracting only part of the information, which was arranged in such a way that it could be readily extracted from the information not required by the simple equipment. Intermediate receivers would, of course, home in on other levels of the hierarchy, but all receivers would have various degrees of intelligence.

A particularly important feature for such a universal digital approach would have to be its ability to withstand multiple processing without deterioration. Whilst this is receiving some attention in present systems, it can hardly be said that they are designed specifically to be immune to impairments introduced by multiple processing. However, this was the basic benefit offered by PCM when it was first introduced 50 years ago and with appropriate research it should be possible to devise standardized bandwidth reduction and expansion techniques which are as immune to multiple passes as PCM is to multiple regeneration.

Thus, in summary, a field frequency independent universal digital TV system would have the following features:

- substantial spatial and temporal resolution for source coding achieved by physical integration of the sensor with the first stage of bandwidth reduction;
- conditional replenishment of picture information data to achieve optimum balance between spatial and temporal resolution commensurate with bandwidths available;
- variable algorithm strategy to produce 'hierarchy' adaptable to application and channel bandwidth;
- hierarchical signal format to serve various levels of studio and receiver applications; and
- ability to withstand multiple processes in series.

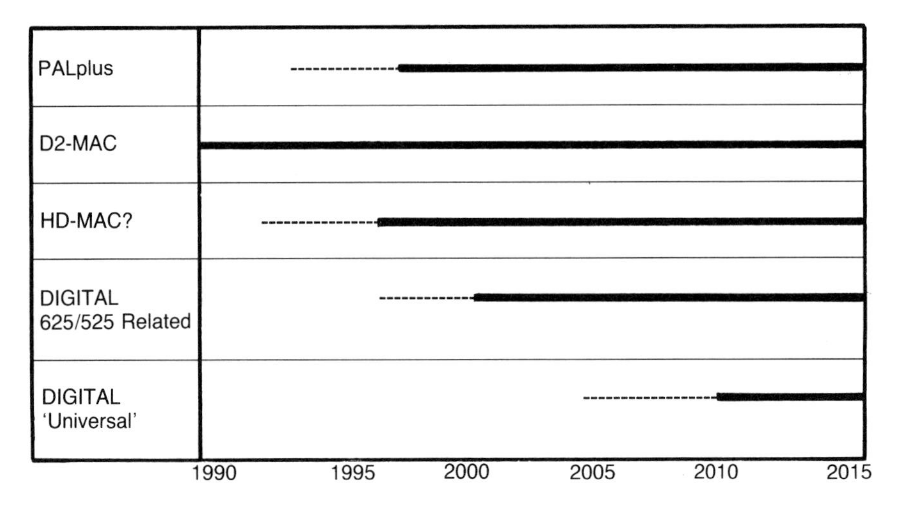

Figure 12.15 Possible timescales for new European TV services.

12.7 CONCLUSION

It is unthinkable that we should continue in the next century to use a large amount of the bandwidth to tell TV receivers every 20 milliseconds in great detail that nothing much has changed during the previous second or so! This will however only be enabled by the non-backward compatible step to digital TV which will become established when the population of analogue receivers has declined significantly.

Figure 12.15 gives my estimate for the possible time scales for new broadcasting services. Sadly a universal digital TV approach may have to wait for a second generation of non-backward compatible digital services perhaps geared to a generally accessible BISDN.

ACKNOWLEDGEMENTS

The author takes pleasure in acknowledging the work of his colleagues in the BBC and in the European Broadcasting Union and thanks the BBC's Director of Engineering for permission to publish.

REFERENCES

ETSI (1992), *European Telecommunication Standard 19 No. 300/163.*

Marsden, R.P., Allen, J.J., Connell, G.J., Lande, J.P. and Mulder, H. (1990) A multi-gigabit optical business communication system using wavelength and time division multiplexing techniques, *16th European Conference on Optical Communications*, September, pp. 779–786.

Pommier, D., Ratliff, P.A. and Meier-Engelen, E. (1990) The convergence of satellite and terrestrial system approaches to digital audio broadcasting with mobile and portable receivers, *EBU Technical Review No. 241/242*, June/August.

Sandbank, C.P. (1980*a*) Communications in the 21st Century *Proc. Instn. elec. Engrs*, Pt A, **127**, (1), January.

Sandbank, C.P. (1980*b*) *Optical Communications*, John Wiley & Sons.

Sandbank, C.P. (1990) *Digital Television*, John Wiley & Sons.

Spillmann and Werner (1990) *Visual Perception – The Neurophysiological Foundations*, Academic Press, pp. 103–128.

Storey, R. (1986) Motion adaptive bandwidth reduction using DATV, *BBC Research Department Report RD 1986/5*.

Thiele, G., Stoll, G. and Link, M. (1988) Low bit-rate coding of high quality audio signals, *EBU Technical Review No. 230*, August.

Thomas, G.A. (1987) Television motion measurement for DATV and other applications, *BBC Research Department Report RD 1987/11*.

13

Optical switching

J.E. Midwinter, FEng, FRS, University College London

Optical fibre technology has made available massive transmission bandwidth and, when coupled with fibre amplifiers, seems likely to provide lossless 'data-pipes' having of order 4000 GHz bandwidth, more than an order of magnitude greater than the whole radio and microwave spectrum! In contrast, electronic logic circuits are typically limited to a few Gbit per second for medium complexity logic yet optical logic is struggling even to come close to the performance of electronics. At the same time, network developments, such as the deployment of asynchronous transfer mode (ATM) systems, are rapidly escalating the functional complexity sought at the switching nodes.

Optical technology has been proposed as the solution to these problems but several initial studies show that the match between the capabilities of optics and the requirements of the emerging network is not good. This paper will thus examine what optics can do, how it can supplement electronics and how it may also lead to a rethinking of the network format in order to localize the difficulties and exploit better the unused optical capability.

13.1 INTRODUCTION

Since the original ideas were floated of using optical fibres for telecommunications transmission (Kao and Hockham, 1966), a revolution has occurred in the subject, starting with the first installations of working systems in about 1980 (Berry *et al.*, 1978; Midwinter and Stern, 1978; Hill, 1980) to the position today whereby optical technology stands supreme for high bit rate cable transmission overland or undersea, between and within cities and increasingly towards or entering business premises or perhaps private homes. The astonishing rate of progress is

Communications After AD2000. Edited by D.E.N. Davies, C. Hilsum and A.W. Rudge. Published in 1993 by Chapman & Hall, London, for The Royal Society. ISBN 0 412 49550 3

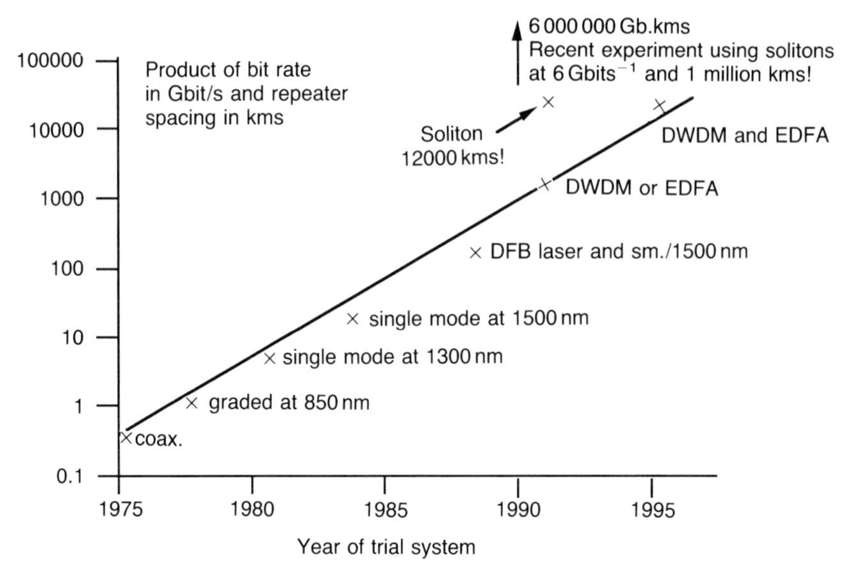

Figure 13.1 Plot of the product of bit rate (in Gbit/s) and achieved repeater spacing in kms versus time.

epitomized best by two summaries and was discussed in my earlier paper on Photonic Switching (Midwinter, 1992). Figure 13.1 shows how the product of achievable repeater spacing and bit rate has changed in the last decade or so, with advances coming at an astonishing rate and growing evidence that, far from flattening out, new advances are occurring even more dramatically as news of long distance soliton propagation studies is announced.

The second milestone, which is included in the plot of Figure 13.1 but only obliquely, concerns the development of the Erbium-doped fibre amplifier (EDFA) (see for example Payne, 1989). This amplifier formed from a few metres of doped fibre, can be sliced in series with the transmission fibre after, say 40 km, to boost the optical signal back to its original level and with the addition only of the fundamental quantum noise associated with every gain process. Figure 13.2 shows the spectral transmission window typical of 50 km of silica fibre at 1500 nm with the EDFA superimposed upon it.

The possibility seems now to exist, not only to operate fibre transmission systems without repeaters across the Atlantic or Pacific but more importantly, to transform the telecommunications network so that the transmission side becomes data transparent in a new sense. With 3000 GHz per fibre of spectrum space available and tuneable semicon-

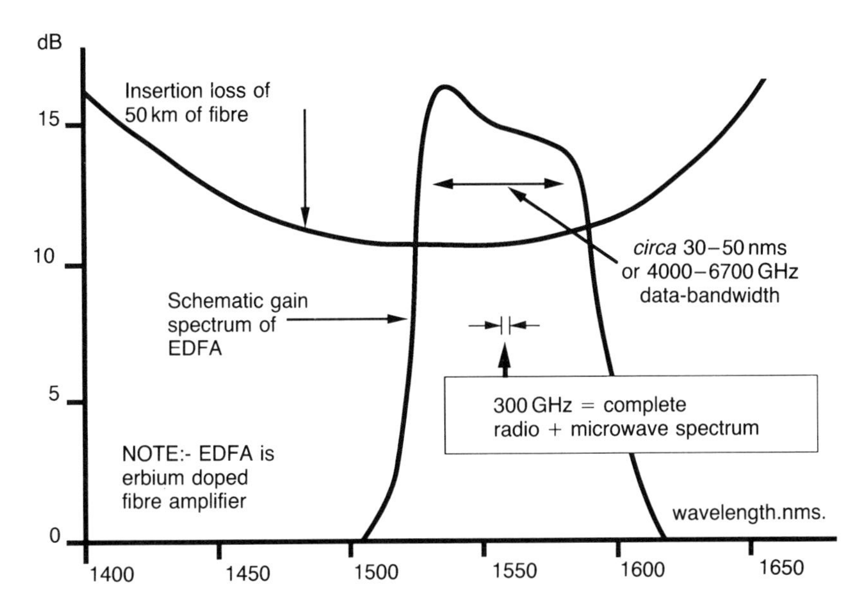

Figure 13.2 The attenuation spectrum of 50 km of typical low attenuation single mode fibre with the gain spectrum typical of an EDFA superimposed showing the potential for a lossless data-pipe with spectral bandwidth of some 3000 GHz or more.

ductors laser that can tune across that band with natural (unmodulated) linewidths of 10–100 MHz, the potential exists to carry vastly greater numbers of channels within the fibre. This comes at a time when the range of services potentially available is increasing to embrace a wide variety of multi-media PC-based communications to supplement video and superior still (colour) picture facsimile etc., all of which could exploit a greater bandwidth capability if it existed. The network is also changing to the synchronous digital hierarchy (SDH) and may well see the introduction of increasing amounts of asynchronous transfer mode (ATM) packet-like traffic included. However, all this implies massive and rapidly growing complexity at the major nodes of such a network, which are already sources of difficulty. Many people have thus wondered if optics can contribute to the solution of this problem.

13.2 OPTICAL ATTRIBUTES RELEVANT TO SWITCHING

Given that switching today is totally dominated by electronics, we may reasonably ask why even contemplate using light. There seem to be three factors that make photonics distinct from electronics.

1. Photons do not interact, except via electrons, whilst electrons interact very directly and strongly. This has a number of results. In long-distance transmission, photons are far superior to electrons in large part for this reason, with the result that the optical signal information has minimal interaction with the medium in which it propagates and with other parallel transmission media. However, if we wish to bring about logical interaction between optical data streams, then we must allow them to interact with electrons in a solid medium in some form, either through the media of (lossless) reactive extremely-fast non-linear interaction (i.e., a soliton switching element) or through (lossy) absorptive reactions such as with the conduction electrons in a semi-conductor optoelectronic device (i.e., an S-SEED to be discussed later). In the former case, high powers tend to be needed and in the latter, the device is effectively an optically triggered electronic device so that the advantage of using optics is somewhat obscured. Despite all the enthusiasm for 'all optical computers', this factor seems to argue against them, other things being equal.
2. The electrical and radio spectrum extends to about 300 GHz upper frequency. The carrier frequency of the light used for optical communications systems is typically 200 000 GHz or 1000 times greater. An immediate result of this is that to carry a given data bandwidth on an optical or electrical carrier fundamentally implies that the fractional bandwidth in the optical case will be 1000 times smaller, with the result that problems of dispersion, timing and data-tagging (by choice of discrete carrier frequency) are all in principle massively reduced in the optical region. It also implies that very much shorter pulses can be handled in the optical domain and this is clearly true, as we observe a developing femtosecond pulse capability and that there is a vast unused spectrum available in the optical region awaiting exploitation along 'radio' lines. This general feature of optics has led to great interest in both the exploitation of 'ultra-fast': techniques to demonstrate picosecond time domain code division multiple access (CDMA) routeing and to a very wide range of dense wavelength division multiplexing (DWDM) systems studies.
3. Intimately linked to 2. above is the result that optical wavelengths are very much shorter (by 1000 fold) than the shortest radio waves and this opens the possibility of establishing free space communications by beams whose area is reduced by λ^2 or 10^6. This has led to a strong interest in the possibility of exploiting imaging optical techniques to handle large parallel arrays of optical beams interconnecting planar arrays of input/output (I/O) devices to generate compact very high performance free-space imaging interconnects.

13.3 THE SWITCHING PROBLEM

Most of the switching problems being discussed here centre around the question of how to handle the appalling complexity that is developing at the logical interface between the customer and the 'intelligent super-versatile network'. This is illustrated in Figure 13.3 which points out the fact that the major functionality of the switching system is concerned with the routeing and supervisory activities, not the actual physical routeing of the message information. Hence this author sees no role for optics in this interface other than perhaps to provide optical interconnects between large electronic processors since the attributes of optics appear to be almost orthogonal to the requirements placed upon the processors at this level. Moreover, it is salutary to note that even when a high data rate stream exists in a telecommunications network, it will normally have been assembled by time multiplexing together a very large number of lower bit rate channels, each of which is destined for a different destination. The switch will thus usually have to treat them as separate entities, not the whole data stream as a single high data rate channel. This further emphasizes the complex trade-off inherent in switching and seen in Figure 13.3.

A partial exception to this reasoning occurs in the high speed part of the core network where there may be a need to move large blocks of multiplexed traffic en-bloc from one bearer to another or where there is a

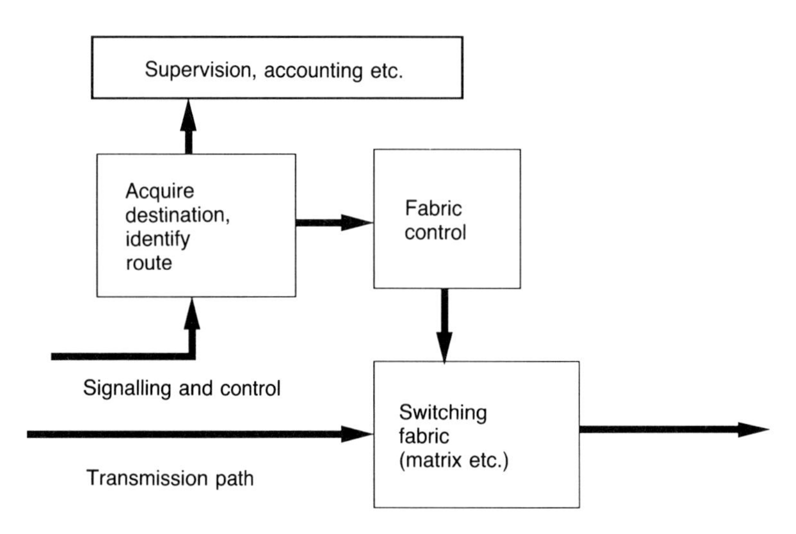

Figure 13.3 Switch contents! Optics seems best placed to impact the switching fabric only and generally this is where the fewest problems exist.

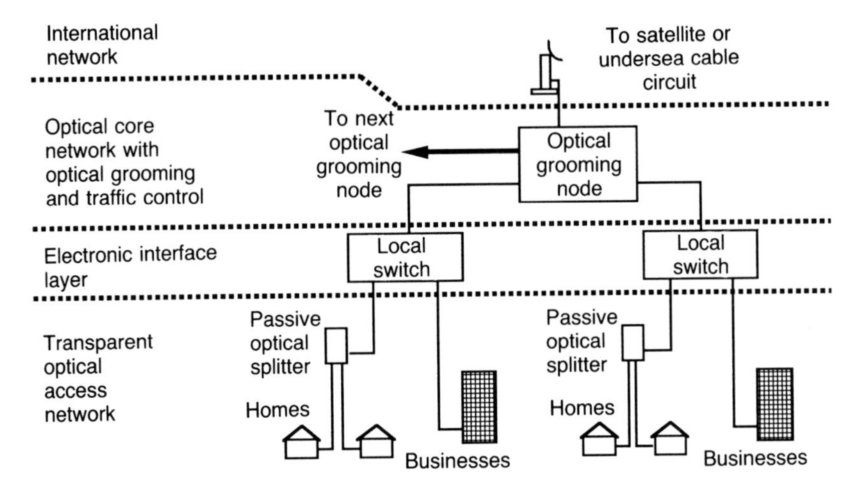

Figure 13.4 The future network will all-optical layers.

need to reconfigure the network to accommodate slowly varying traffic flow changes linked to changing network useage patterns during the working and recreational day. There is also a need to be able to reroute traffic rapidly when a JCB cuts through a major transmission system cable or when a system has to be taken out of traffic carrying service for maintenance purposes.

Thus, there do appear to be possibilities for exploiting optical routeing techniques on either side of that interface although to most switching people this might not be considered 'switching' but rather 'intelligent transmission', as illustrated in the Figure 13.4.

What we are proposing here is that in order to exploit fully the proven transmission capability of optical technology, it might soon start to become attractive to plan a high speed core-network that would for practical purposes be optically transparent and a local access network that would also be so, interfaced through a notionally single electronic layer. Ideas on the possible form of such networks will emerge from our discussion below.

Another area in which 'optical switching' might find application is the much more limited one of the closed multi-terminal network of the general schematic form shown in Figure 13.5 which might represent a super Fibre Distributed Digital Interface (FDDI) local area network (LAN) operating at 1–10 Gbit s^{-1} throughput of the communications core of a large multi-processor computer where the data flow rates might exceed 100 Gbit s^{-1}.

Thus we will not propose that there is any significant likelihood that

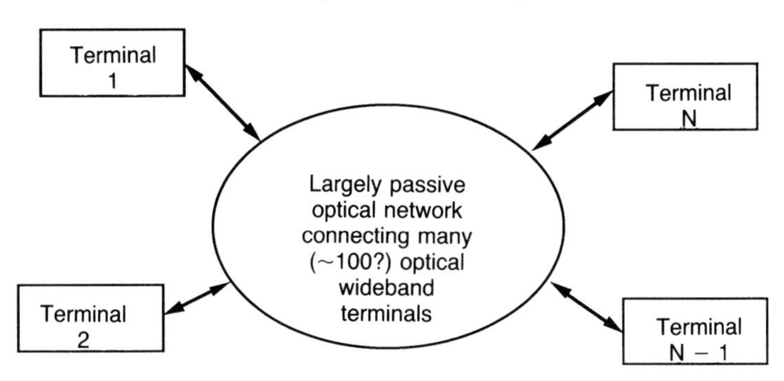

Figure 13.5 Closed multi-terminal network amenable to all-optical implementation to provide extremely high data throughout in a non-blocking manner.

optics will displace electronics from switching, any more than we would suggest that it would do so in computing but we will suggest that there are some interesting opportunities to exploit techniques within a variety of networks that lead to major performance gains.

Finally we should note that a fundamental problem involved in all attempts to exploit the bandwidth of optical carriers can be summed up in the form shown in Figure 13.6.

We find that digital electronics in complex circuit form rapidly runs into difficulty as 1 Gbit s^{-1} clock rates are approached and only the very simplest circuits operate above that level because of the huge problems in maintaining accurate timing etc. at these rates. On the other hand, at these rates one is barely scratching the surface of the available optical bandwidth. At present, the automatic design tendency is to multiplex

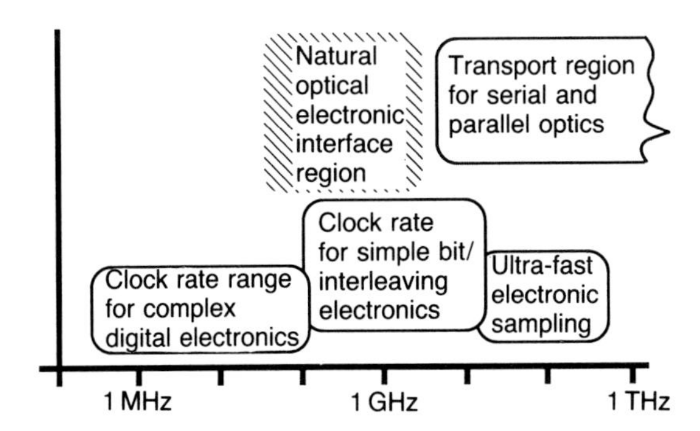

Figure 13.6 The interface between electronics and optics which tends to restrict the use of optical bandwidth to blocks that are convenient to handle electronically.

electronically to the highest level possible electronically and then to transfer into the optical domain. We will suggest later that this may not be the best route if the capability of optics is to be properly exploited.

13.4 EXAMPLES OF OPTICAL SWITCHING TECHNOLOGY?

13.4.1 Electro-optic switching matrices

The technology that most people think of when discussing optical switching is that of integrated optics based upon electro-optic materials like lithium niobate. Using these materials, simple 2 × 2 couplers can be made that can be electrically controlled to vary the connections between the two single mode inputs and the two output ports from exchange to bypass, for example, see Schmidt and Alferness (1979). The devices are formed by diffusing the single mode guide structure into the crystal surface using a lithographic process to define their position and dimensions. In this way, the single mode guides are just buried within the electro-optic crystal. Control electrodes are applied externally to which the application of ~10 V signal suffices to change the state of the device. Small arrays of these can then be formed on a single crystal to build up a small switching matrix. Typical reported sizes are 8 × 8, 1 × 16 and 16 × 16. Larger matrices are not possible because the maximum crystal dimension available is about 10 cm and the individual devices are typically spaced by a centimetre once allowance is made to accommodate the interconnection guide structures which also has to be diffused into the crystal. The individual cross-points can switch in sub-nanosecond times but the problems of electrically driving even an 8 × 8 matrix are likely to slow that in practice to tens of nanoseconds.

An example of a *tour de force* constructed using this technology was recently reported by NEC (Burke *et al.*, 1991). It demonstrates both the strength and weakness of the approach. The 128 × 128 matrix built up using many discrete but smaller cross-point arrays is illustrated in Figures 13.7 and 13.8 and the vital statistics for the array are as follows.

Active component count

Number of 4 × 7 switch matrix modules: 176
Number of 8 × 8 switch matrix modules: 49
Number of directional couplers in each 4 × 7 stage: 28
Number of directional couplers in each 8 × 8 stage: 64
Number of 4 × 4 building blocks: 548
Total directional coupler count: 8064
Number of laser amplifiers: 784

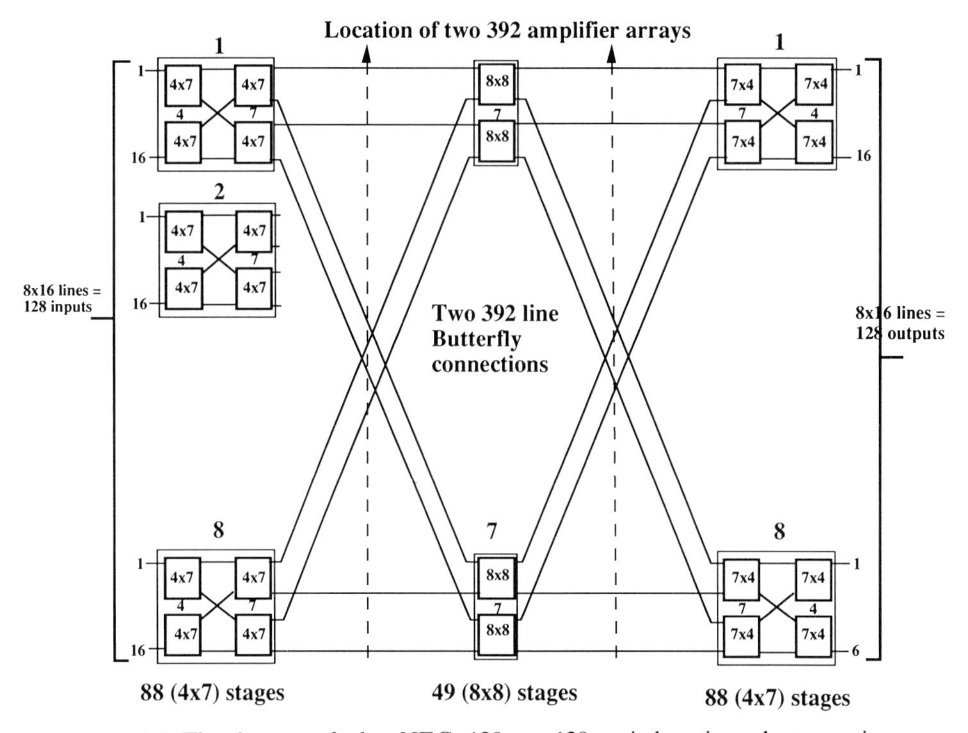

Figure 13.7 The layout of the NEC 128 × 128 switch using electro-optic crosspoint arrays as its building blocks.

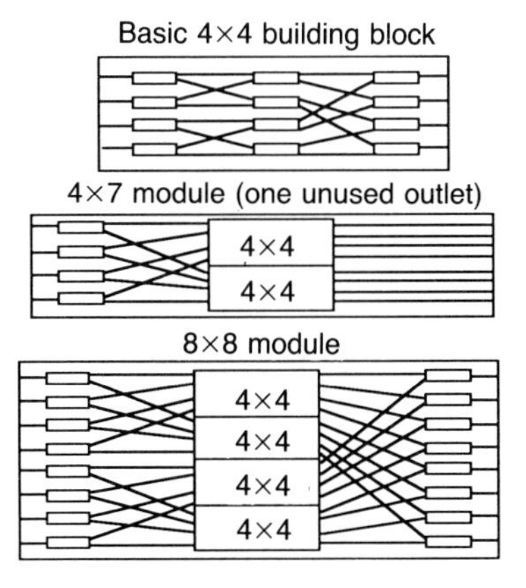

Figure 13.8 The constituent components of the switch matrix of Figure 13.7 showing the discrete switch structure.

Passive (single mode) connection count

Interconnects/butterfly (incl.Amps): 1568
Guided wave line in 4 × 7 stages: 224
Total number of guided wave connects: 4288
Optical insertion loss without amplifiers: 48 dB

By optical standards, the component count in this assembly would be enormous if it were to be fully equipped although presumably, once assembled, there is good reason to believe it would function. With 8064 discrete directional couplers, the electrical drive is fairly awesome particularly if the matrix needs to be reset rapidly to handle time multiplexed data. One also recognizes that the insertion loss of the assembly (48 dB) is such that two chains of amplifiers are necessary to boost the signal levels back to an acceptable level and these amplifiers are inserted into the matrix as indicated, further complicating the overall assembly. Complexity of this level is taken for granted in electronics where the component technology and circuit board development allows it to be implemented with ease, but using single mode optical technology, it remains a *tour de force*. To what extent this will change as the technology matures remains an interesting subject for debate.

On the speed side, it must be doubted whether this technology can ever hope to switch (reset) as fast as an all-electronic implementation so that it seems fundamentally unsuited to operating directly upon the bytes within a time multiplexed data stream. However, since each channel, when established, can handle a large bandwidth (many THz?) applications may be found where relatively slow routeing is required for wideband data channels. Protection switching might be an example.

13.4.2 Semiconductor guided wave switches

Another family of similar devices might be based upon epitaxially grown waveguides in III–V semiconductor materials. A variety of techniques have been proposed ranging from using semiconductor laser type structures as electrically controlled amplifiers/attenuators to switch signals through to using changes in the refractive index arising from carrier injection (Kawachi, 1987) or the changes in absorption edge position in a reverse biased MQW structure to produce switching between adjacent guides in an analogous manner to the oxide crystal electro-optic materials described above. The advantage of using these materials is that the effective electro-optic effects that can be accessed are much larger so that very much shorter switch structures can be achieved, promising much higher device packing densities. In addition, the possibility of achieving gain in the same structure is also attractive as a means of achieving '0 dB insertion loss' switches. However, difficult problems exist in such structures in achieving low enough scattering and cross-talk losses and despite the initial promise, large-scale switch structures have not so far been reported with experimental devices limited to being of order 4 × 4. Such switches would also be subject to a similar problem regarding route set-up time as the lithium niobate time, although if they should prove to be physically smaller, then the problem might well be more readily solved.

13.4.3 Wavelength division multiplexed routeing

A completely different guided-wave approach to routeing signals in optical networks is to exploit the optical spectrum availability by assigning a different optical carrier wavelength to each channel that is to be routed. We have already noted that a huge spectrum is available, typically of order 3000 GHz just within the EDFA window. Thus, if we were to assign a GHz of spectrum to each channel, one might conclude that 3000 wideband channels could be handled in parallel simultaneously through a single fibre and since each would be tagged by a discrete carrier wavelength, it could in principle be identified and routed entirely within the optical system by optical means alone.

In practice, life is more difficult. In Figure 13.9, we show the typical tuning range and linewidths for a variety of semiconductor laser sources.

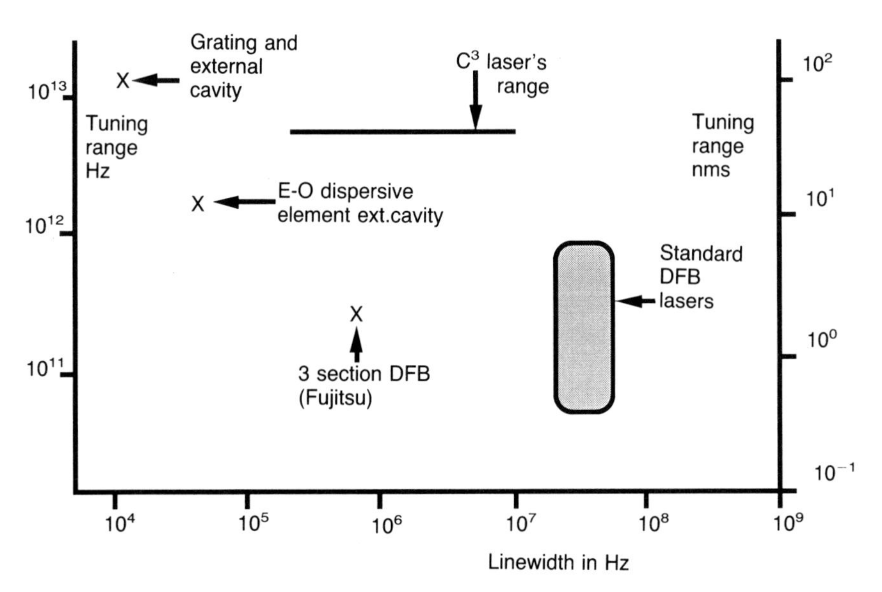

Figure 13.9 Plot of typical linewidths and tuning ranges for a variety of different semiconductor laser sources.

We see that the standard DFB laser with suitable electronic control is tuneable over a frequency spectrum of about 100 GHz so in practice the number of channels it could access is likely to be less than 100 and typically values nearer to 30 seem to be considered. However, Figure 13.9 also shows that with more sophisticated dispersive cavity structures, semiconductor sources can simultaneously achieve narrower natural linewidths and much greater tuning ranges so a degree of optimism should be felt for the principle.

Having launched information into a network suitably optically coded by carrier frequency or wavelength, the receiver requires a tuneable filter of similar linewidth in order to extract the message destined for it and it alone. The general format of such networks looks like that of Figure 13.5 with the central passive power splitter serving to split the power from each transmitter equally among all receivers. Thus if there are N terminals, in principle $1/(N - 1)$ of the power arrives at each receiver (since terminals do not need to communicate with themselves). Given that a typical link might tolerate upwards of 30–35 dB insertion loss, values of N approaching 1000 are feasible in principle in compact networks and the actual values are more likely to be controlled by the tuning properties of the source and receiver filters.

Many filter types have been proposed for this type of network. Generally one may consider fixed or tuneable filters, depending upon the channel routeing technique to be followed (tuneable source, tuneable receiver or both). The commonest element in optics to carry out narrow spectral band filtering is the Fabry–Perot etalon filter which corresponds to the high-Q cavity resonator of microwaves. The optical filter of this type is characterized by having two high reflectivity mirrors spaced by a distance L, which if mounted on a material of refractive index n gives rise to transmission peaks spaced every $d\lambda = \lambda^2/2nL$. Wavelengths between these peaks can then be selected by electronically or mechanically changing the spacing L or the index n. This is known as the free spectral range (FSR). The finesse (F) of the filter is the FSR divided by the full width half maximum (FWHM) of the filter response – F = FSR/FWHM which is a function of the mirror reflectivity. For the Fabry–Perot filter, $F = \pi R/[2(1 - R)]$ so for $R = 99\%$ we find $F = 155$ implying that about 155 spectral bands could be resolved. However, for practical communications systems, the actual number that can be handled is considerably fewer, perhaps as few as 1/3 to 1/10 of the finesse, F. In practice, of order 30 channels can be achieved today. For example, 0.17 nm resolution (FWHM) has been reported over a 50 nm FSR in a liquid crystal tuned device.

Another class of filters is based upon two beam interferometers, typically of the Mach–Zehnder type. These offer low resolution wavelength splitting with the coupling between any one input port and one of two output ports varying roughly sinusoidally with wavelength and with peak transmission again occurring at wavelengths set by a relationship of the form $\lambda + N\mathrm{d}\lambda$. where N is an integer and $\mathrm{d}\lambda$ is controlled again by the path difference between the two arms. Multiple stages of such elements can be cascaded together splitting successively smaller ranges so that a finely spaced comb of wavelengths can in principle be split by a series of two ways splits into a single line into a single output port, with a resolution of 100 channels at 10 GHz spacing reported.

Finally, one should note that using an optical local oscillator and optical heterodyne detection, it is possible to scan as much of the optical spectrum as the source will tune across with the spectral resolution set by the IF bandwidth of the electrical receiver. However, the penalty of this approach is a rather complex receiver which militates against its use in many applications. At present the position is thus that many interesting and very powerful networks can be built but networks having the full potential apparently on offer from optical technology are either extremely expensive and difficult to assemble or are otherwise impractical.

However, it should be noted that networks of this type offer the huge advantage of being fully non-blocking and offering full bandwidth communication between every pair of terminals on a continuous rather than time-shared basis. Hence, throughputs of $1\,\mathrm{Gbit\,s^{-1}}$ per terminal on a 32-terminal network are probably just about possible now and this might be extended to more terminals and higher bits rates within the next few years. Such switched networks are roughly equivalent to today's LAN or FDDI networks but with vastly greater data throughput rates which are achieved without the need for immensely complex electronic interfaces. However, they do little to solve the control overhead problems already referred to which become particularly acute when handling ATM traffic (Brackett, 1991). Note that FDDI operates with a single $100\,\mathrm{Mbit\,s^{-1}}$ time-shared channel versus the WDM network with perhaps $32 \times 1\,\mathrm{Gbit\,s^{-1}}$, an increase in throughput of at least 320!

13.4.4 Ultra-fast switching and CDMA

An alternative approach to exploiting the spectral bandwidth of optics is to move to the ultra-fast regime, in which one attempts to use the ability inherent in optical systems to transport and process in simple interactions picosecond or femtosecond pulses. Two broad approaches can be discerned.

One involves the uses of solitons or other ultra-fast pulses to make possible extremely fast logical interaction between data streams in non-linear optical gates, typically formed from coils of optical fibres which at high power levels exhibit a very fast non-linear intensity-dependent refractive-index (Friberg *et al.*, 1988). Whilst it is possible using such technology to effectively add or drop extremely fast pulses from a serial stream and to perform logical *AND* or *OR* operations in very simple circuits, the system issues of how such elements could conceivably be used to advantage in an optical network remain unclear where the normal requirement is to perform a more complex processing operation, such as is implied by header routeing of packets.

The second approach has been to draw upon ideas developed in radar systems for extracting signals from noise using transmitted signals coded with pseudo-random sequences with filters matched to those codes in the receivers (Weiner *et al.*, 1989). The optical equivalent of this has been to code ultra-fast pulses by splitting them into their spectral components, imposing upon the spectrum some random transmission filter that in effect converts the fast pulse into a longer time-coded pulse sequence and then to use the inverse spectral filter to recover the signal from the mixture emerging from the fibre which contains both the desired pulse stream and streams from other transmitters aiming for other destinations. In this situation, each data pulse is given a code that can be recognized by the destination receiver. Thus each data pulse is stretched into a coded pulse sequence so that once again, bandwidth is used very wastefully to achieve routeing. The initial indications seem to be that while this has led to extremely elegant optical experimentation, it is a less effective technique for routeing by using the available spectrum space than the WDM technique.

13.4.5 MQW planar free-space-addressed switching arrays

A completely different approach to the switching and routeing of light has grown out of early thinking on parallel-free-space interconnected optical logic processing for 'optical computing'. The most successful devices for this have unquestionably been the Self Electro-optic Device (SEED) arrays fabricated in GaAlAs MQW materials (Miller *et al.*, 1985). Their structure is shown schematically in Figure 13.10.

The PIN diode structure is grown using MBE or MOCVD on top of an epitaxial multilayer-dielectric-reflector to form a normally-addressed reflecting device. By changing the voltage across the structure, the absorption edge can be shifted so that the effective absorption close to it changes and with suitable design, this can result in substantial changes in the reflectivity, e.g., 20 dB form a low value of 3 dB insertion loss (Whitehead *et al.*, 1989). In this form the device looks very promising for use as an opto-electronic interface device to be used for taking data from the electronic domain and impressing on the optical domain or vice-versa,

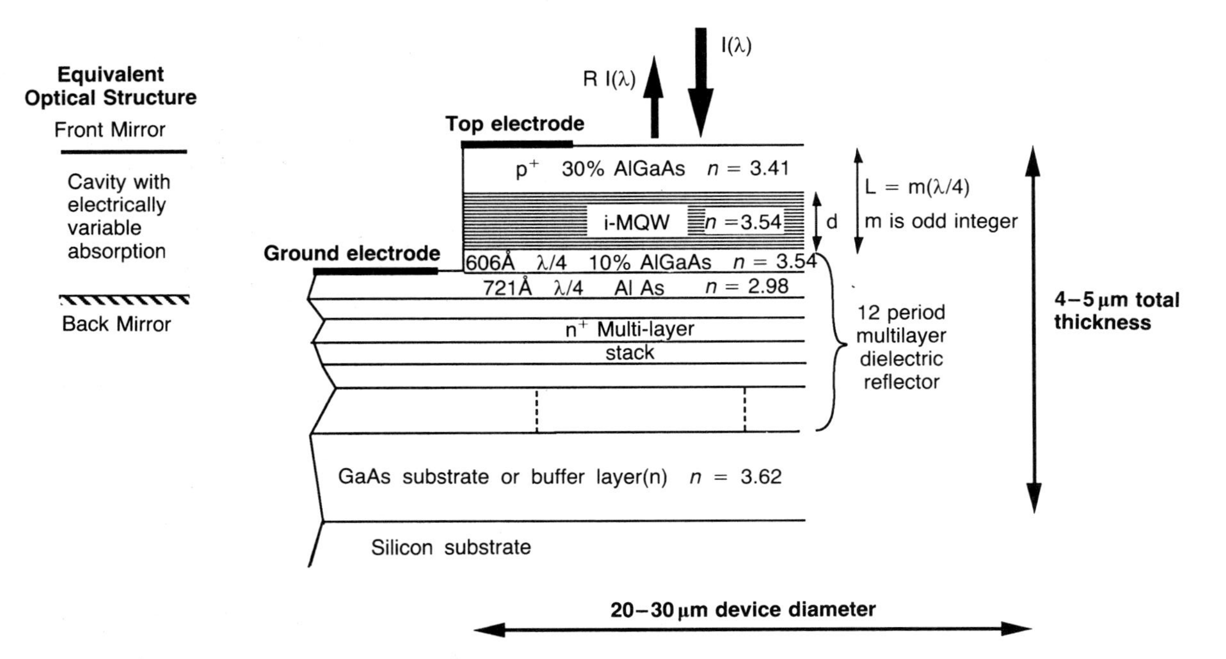

Figure 13.10 Typical structure of a MQW modulation or SEED device.

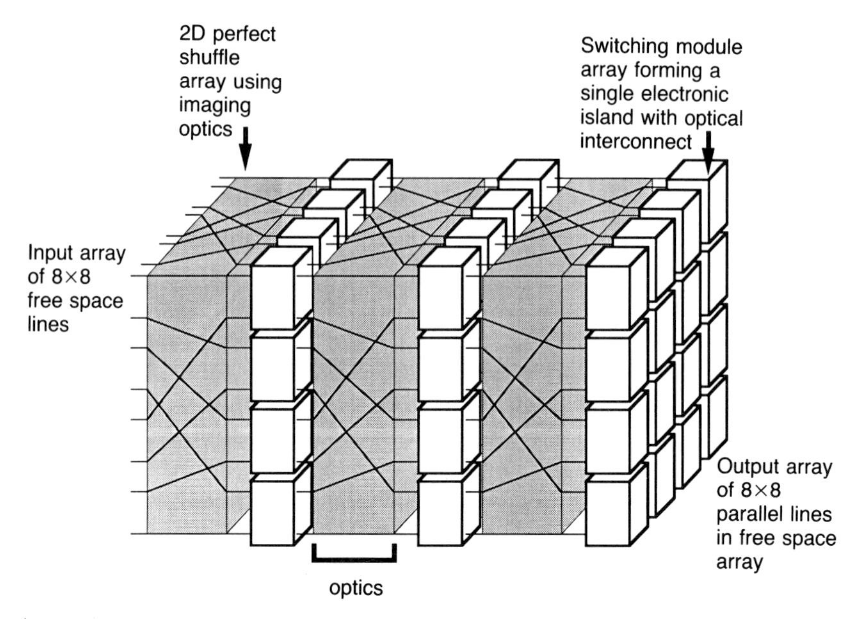

Figure 13.11 Schematic of an optical switch with the matrix logic based upon optical and/or electronic logic interconnected using free space parallel imaging objects.

since the same device can operate as a powerful photo-detector. Thus, small arrays, say of 8 × 8 dimension, may well serve as very high performance pin-outs for powerful processing chips.

The light that is absorbed by such a device generates photo-current and this can be used in the external circuit to couple several devices together to form a single unit of higher functionality. For example, two such devices in series form the symmetric SEED (S-SEED) which has been used as an optically-addressed self-latching modulator. Large arrays can be produced (e.g., 256 × 128 S-SEEDs) which can be linked with imaging optics to form optically-controlled spatial light modulators (SLMs) capable of fairly rapid (msec to nsec) switching depending upon the optical power level. The attraction of such elements is that they only require electrical power connections, all their data connections being carried out in the optical domain. This has led to speculation on the possibility of constructing complex switching matrices using free space interconnections with imaging optics, shown schematically in Figure 13.11.

The group at ATT-Bell Labs at Naperville (Cloonan *et al.*, 1990; McCormick *et al.*, 1990) have demonstrated some extremely impressive results using this technology but it remains unclear whether it offers serious competition to all electronic implementations of the similar switching structures. There is little doubt that large free-space arrays of such devices offer a massive data-through-put capability but whether they offer any comparable advantage from the control point of view is much less clear. What has emerged clearly from these and related studies (Feldman *et al.*, 1988; Midwinter, 1990) is the fact data flow onto and off from an electronic chip suffers from serious 'pin-out' bottleneck problems and free space optics may well have a role to play here, so that we find a rapidly growing interest in trying to combine the strengths of the free-space optical interconnect approach with those of electronic processing in some form of hybrid technology.

13.4.6 Optical interconnects

Strictly, interconnects between electronic processors should perhaps not be considered under 'optical switching' but it is so clear that 'optics within switching' is growing in importance for its interconnect capability, we cannot ignore it. Moreover, in almost any application as we have seen above, there is an interplay between optics and electronics so that the position at which the boundary between the two technologies is drawn becomes increasingly blurred.

Optical technology primarily in fibre form offers great potential for use in place of metal cable for inter-rack and rack back-plane interconnection and many manufacturers are tackling the engineering problems these applications raise. At circuit board level, to interconnect chips using guided-wave optics remains a tough challenge if it is to be achieved cost effectively. Optical printed circuit boards exploiting silica-guides on silicon mother-boards look attractive for many purposes since silicon is a stable and robust material that lends itself to micro-fabrication techniques using lithography. Thus precision locating slots are readily formed on silicon to locate electronic or optoelectronic chips and to interface them to planar waveguides formed on its surface. However, it is not obviously a low cost technology, the wiring patterns are single layer and waveguide cross-overs are difficult to fabricate and the O-E interfaces tend to be heavily power consumptive since they will usually involve long bond wires and significant stray capacitances. Moving down to the chip-carrier level, optical free-space interconnects may find application within the carrier but much work remains to be done to establish a viable technology to do that with work spanning both hybrid and monolithic approaches to a solution (Barnes *et al.*, 1989).

13.5 SYSTEM IMPLICATIONS

It is fairly clear from the above discussion that most of the optical switches that have been described, where the data path is purely optical, offer extremely high data bandwidth once a connection is made, just as does the optical fibre itself. The route set-up time is likely to be dominated by the time taken for the electronic control system to identify a suitable route and generate the local control signals necessary to set up the path through the optical switching fabric. This is likely to be very much greater than the minimum bit interval that can be transmitted through the fabric and this has led to proposals for optical time-slot-interchange switches in which very high data rate message sequences are interspersed with empty time slots or guard bands to allow time for switch resetting (Thompson and Giordano, 1987). In addition, the likely complexity of the fabric is limited, with most optical techniques at present struggling to extend to 100 × 100 sizes although as we saw, the dense WDM technology does offer a possibility of perhaps 1000 channels.

These general conclusions are summarized in Figure 13.12 which also shows some other potential options. The SEED-free-space-array route might possibly extend to high complexity since that has always been claimed the greatest attribute of free-space imaging interconnects although the engineering problems remain severe. In addition, we have shown the very low complexity but high-speed option that exists in optics to handle a very few logical channels at very high data rates.

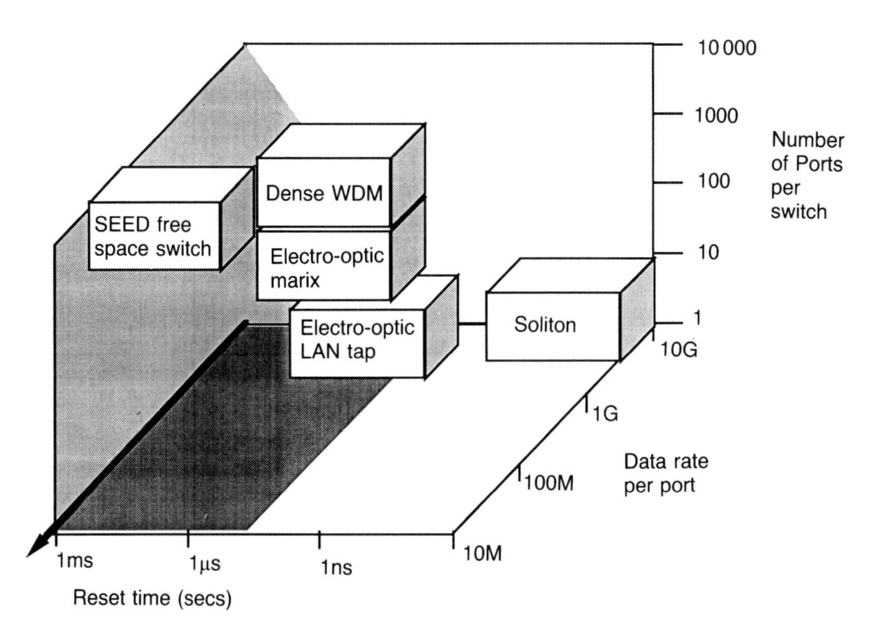

Figure 13.12 3D plot indicating the relationship between the various optical switching technologies and their performance parameters.

If we now examine the networking problems again, we see that optics seems extremely ill-matched to handle the problems of ISDN circuit switching, where very large numbers of low bandwidth channels have to be routed. However, if we look to the main network of the switch, then there may be interesting options for using optical technology to enhance the flexibility and power of optical transmission systems, for example by offering an all-optical 'add-drop multiplexer' capability to a high speed transmission system at an otherwise transparent node. This is very similar in concept to the much discussed requirement for a digital cross-connect capability in the USA whereby the overall traffic capability of the high-speed core of the telecommunications network can be slowly reconfigured to match changing traffic demands during the course of the day.

Referring back to Figure 13.4, the 'all-optical core network' must be able to handle the re-routeing of major traffic flows in response to both changing demand, equipment maintenance and repair and hardware and traffic growth. How might this be done?

The thinking in virtually all network planning involving optics to date has been to multiplex electronically to the highest level feasible in order to take best advantage of the optical transmission capacity. A corollary of this is that the data rate per optical bearer is very high and is probably an inconveniently large block for traffic grooming or re-routeing purposes, implying that it will always be necessary to return to the electronic domain and to demultiplex several through layers of the hierarchy before any traffic can be split off and redirected.

We suggest that this approach will be made inappropriate by the steady growth of capability to handle wavelength multiplexed channels. Let us indulge in a little pipe dreaming. From Figure 13.6, we can make two points. The first is that a massive mismatch exists between the bandwidth readily accessible in the electronic and optical domains and second is that by exploiting many optical carriers, the bandwidth available can be filled using readily handled electronic channels such as the STM-1 or STM-4 levels rather than always planning to enter the optical spectrum via an STM-16 multiplex. Electronic thinking has always sought to push to the highest time division multiplex before entering the optical domain. We suggest this may well be wrong.

Suppose for thinking purposes the WDM technology advances to the point where 1000 optical carriers are accessible. What implications might this have? The first observation we can make is that a single fibre can now carry a huge capacity without having to assign greater than an STM-1 level to a given carrier. What values might be appropriate?

We have seen that the fibre and EDFA transmission window seems to be about 30 nm wide. This corresponds to a spectrum space of 4000 GHz. Accordingly, we might consider our 1000 wavelengths comfortably set at

2 GHz spacings within the window. If we chose STM-1 as the basic data level at which to enter the optical regime, then 100 channels at STM-1 would correspond to a data rate 15.5 Gbit s^{-1} per fibre and there would still be a massive choice of 10 wavelengths for each desired channel. With such a low utilization rate for wavelength channels, the problems of routeing through a network of UK dimensions becomes almost trivial since at most one might wish to traverse perhaps 6–8 major nodes on any route. There would thus always be unused wavelengths that could be selected and that would route transparently through from source to destination.

However, to implement such a system implies some very sophisticated integrated tuneable transmitter systems, offering perhaps all 100 channels from a single opto-electronic integrated circuit (OEIC) chip. Also inherent in such a proposal are tuneable optical filters that could select and redirect the chosen channels at each node. Such filters would be on the limits of our technology today. There is also presupposed a transparent optical switching matrix that can handle the traffic redirection implications of such a network at a major node and it seems reasonably clear that this would need to be of the wavelength-space-wavelength type of switch, by analogy with the time-space-time switch currently used in electronics. Finally, such a network implies the existence of optical frequency standards that would allow such a comb of wavelengths to be established and operated throughout the UK or any other network operators territory. These do not appear to be insurmountable problems but they do represent a considerable challenge.

The overwhelming attraction of this approach is that, for the first time, it tags a data channel in the optical domain with an optically readable label (wavelength) that allows it to be recognized, accessed and processed without electronic intervention. The network that begins to emerge has immense versatility and is based upon a mass produced set of common building blocks that would be equally applicable to the wideband local network and to high performance LANs. The downside is clearly that at present, the technology it envisages is regarded as experimentally difficult to fabricate, handle and control as well as being expensive, assuming that it is available at all. It thus presents a major challenge to the developers of the OEIC technology to find ways of fabricating *en masse* such sophisticated components.

Given such a model, we are rapidly led to think in terms of a new type of wideband switch which would be of the wavelength-space-wavelength type, analogous to time-space-time switches in electronics but exploiting the new dimension in optics. Such a node then takes the schematic form shown in Figure 13.13.

The role of the wavelength switch is to select from within the wavelength multiplexed group of channels, sub-groups of wavelengths that all

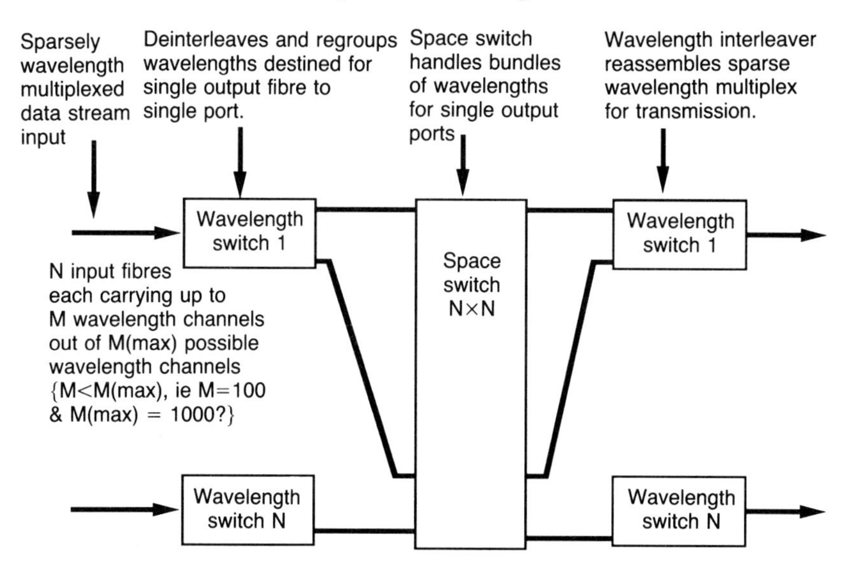

Figure 13.13 Wavelength-space-wavelength schematic to exploit both the transmission potential of optical fibres and the component capability now emerging for use at the node.

wish to depart via a single exit fibre port, and to group those onto a single exit port of the wavelength switch. The space switch then delivers to the exit wavelength switch all wavelength channels destined for that exit port and the switch reinterleaves them prior to transmission. By exploiting the potential for substantial wavelength redundancy, we believe that wavelength shifting at the node can be avoided even within a very substantial network, so that it becomes a truly transparent node whilst retaining a very degree of functionality. It seems reasonable to propose that such a complex switch and the supporting components should be realizable in the laboratory within five years given the present rate of technology advance. The way will then be wide open to plan and implement highly flexible transparent broadband core networks.

13.6 SUMMARY

It seems clear that optical technology does not simply map onto and replace electronic technology but that to be exploited, must be moulded to the problem in such a way that its strengths are exploited and weaknesses avoided or supplemented by alternatives. In general, optics and electronics are complementary to one another and thus optimum solutions often involve mixtures. Optics seems to offer little potential for logical interaction since such interactions can only be produced in (pseudo) electronic devices. The exception is perhaps in the ultra-fast femtosecond time regime although what system applications are likely to arise there is more difficult to see.

The short wavelength of optics opens the intriguing possibility of using free-space imaging interconnect technology to connect planar arrays of devices in switch-like structures. Very impressive results have been reported, massive data throughputs are possible but it is not obvious that the technology solves the difficult control problems inherent in large switches.

The high carrier frequency of optics opens up as yet largely unexplored potential for DWDM systems for which the component technology is as yet inadequate. However, the theoretical possibilities for establishing truly transparent yet highly versatile wideband networks are immense although to some extent if they are to be exploited, they also require a degree of rethinking in network design. The same technology also has immediate application in high-throughput multi-processor computer networks and, once deployed, wideband local networks, so that there is a potentially large market for a range of advanced 'building block' components designed to implement DWDM systems.

REFERENCES

Barnes, P., Zouganelli, P., Rivers, A., Whitehead, M., Parry, G., Woodbridge, K. and Roberts, C. (1989) A GaAs/AlGaAs MQW modulator using a multilayer stack grown on Si substrate, *Electronics Letters*, **25**, 995.

Berry, R.W., Brace, D.J. and Ravenscroft, I.A. (1978) I.E.E.E. Trans. Comm. **COM-26**, 1020–7.

Brackett, C.A. (1991) Capacity of multi-wavelength optical-star packet switches and implications for packet length, *Topical Meeting on Photonic Switching*, Salt Lake City, Utah, 6–8 March, Optical Society of America, Washington.

Burke, C., Fujiwara, M., Yamaguchi, M., Nishimoto, H. and Honmou, H. (1991) Studies on a 128 line photonic space division switching network using lithium niobate switch matrices and optical amplifier, *Topical Meeting on Photonic Switching Paper FA-4*, Salt Lake City, Utah, March 1991, Optical Society of America.

Cloonan, T.J., Herron, M.J., Tooley, F.A.P., Richards, G.W., McCormick, F.B., Kerbis, E., Brubaker, J.L. and Lentine, A.L. (1990) An all-optical implementation of a 3D crossover switching network, *I.E.E.E. Photonics Technology Letters*, **2**, 438.

Feldman, M.R., Esener, S.C., Guest, C.C. and Lee, S.H. (1988) Comparison between optical and electrical interconnects based on power and speed considerations, *App. opt.*, **27**, 1742–51.

Friberg, S.R., Weiner, A.M., Silberberg, Y., Sfez, B.G. and Smith, P.S. (1988) Femtosecond switching in a dual core fibre non-linear coupler, *Optics Letters*, **13**, 904–6, October.

Hill, D.R. (1980) 140 Mbit s^{-1} optical fibre demonstration system in *Optical Fibre Communication Systems*, (ed. C.P. Sandbank) Wiley.

Kao, C.K. and Hockham, G. (1966) *Proc. I.E.E.*, **113**, 1151.

Kawachi, M. (1987) Integrated micro-optic components based on silica waveguides, *European Conference on Optical Communication, ECOC-87*, Helsinki, Finland, 13–17 September.

McCormick, F.B., Tooley, F.A.P., Cloonan, T.J., Brubaker, J.L., Lentine, A.L., Hinterlong, S.J. and Herron, M.J. (1990) A digital free space photonic switching network demonstration using S-SEEDs, *CLEO 1990, Technical Digest Series*, **7**, post deadline paper CPD-1, Optical Society of America, Washington.

Midwinter, J.E. (1990) Communications, VLSI, optoelectronics and self-routeing switches, Prestige Poster Session, XIII International Switching Symposium, Stockholm, May 27–June 1.

Midwinter, J.E. (1992) Photonics in switching, the next 25 years of optical communications, to be published. Proc. IEE. Pt. J, Optoelectronics.

Midwinter, J.E. and Stern, J.R. (1978) Propagation studies of 40 km of graded index fibre installed in cable in an operational duct route, *I.E.E.E. Trans. Comm.*, **COM-26**, 1015–20.

Miller, D.A.B., Chemla, D.S., Damen, T.C. *et al.* (1985) The Quantum Well Self Electro-Optic; bistability, oscillation and self-linearized modulation, IEEE J. Quantum Electron. Vol. QE-21, 1462–76.

Payne, D.N. (1989) Advances in active fibres, invited review paper 20.A-3.1, *Integrated Optics and Optical Fibre Communication IOOC-89*, July 18–21, Kobe Japan, IEICE Japan.

Schmidt, R.V. and Alferness, R.C. (1979) Directional Coupler Switches, Modulators & Filters using alternating $\Delta\beta$ techniques, *I.E.E.E. Trans. Circuits Systems*, **CAS-26**, 1099–108.

Thompson, R.A. and Giordano, P.P. (1987) An experimental photonic time-slot interchanger using optical fibres as re-entrant delay line memories, *I.E.E.E. J. Lightwave Tech.*, **LT-5**, 154–162.

Weiner, A.M., Salehi, J.A., Heritage, J.P. and Stern, M. (1989) Encoding and decoding of femtosecond pulses for CDMA, *Topical Meeting on Photonic Switching, Salt Lake City, March 1–3, Paper ThA-1*, Pub. Optical Society of America.

Whitehead, M., Rivers, A., Parry, G., Roberts, J.S. and Button, C. (1989) Low voltage multiple quantum well modulator with on-off raionbetter than 100:1, *Electronics Letters*, **25**, 984–5.

14

Broadband fibre

T.R. Rowbotham, Director Network Technology, BT

14.1 INTRODUCTION

In the future the availability of inexpensive broadband fibre communications will change everyone's lives. For example, at the press of a button, or on voice command, one of the windows of the office could change to a large colour high-definition flat TV (HDTV) screen. This screen could be used as a videoconference terminal linking the office to colleagues who are working from home.

The team need not be in the same town, or even in the same country, because international calls could be extremely cheap even for broadband services like HDTV.

The office desk could also use advanced technology, the whole surface being a large liquid crystal display, initially displaying a beautiful rosewood veneer. The liquid crystal display could then change to show the pages of documents. The pages could be turned by rubbing the corners. Pages could be separated and moved around the desk by touching then with your finger. The documents could contain not only text and pictures but also speech and moving images. Documents could be more like TV documentaries than traditional paperwork. They could also be seen as projected virtual 3D images. From the desk, databases all around the world can be accessed, including the company's ones, which might be located in the highlands of Scotland to avoid high London rates.

When speaking to foreign colleagues on the HDTV videoscreen, a simultaneous translator could be selected using a centrally provided parallel optical computer. People will be freed of the barriers of language and location, and could deal with whomever they please with ease. In the desk there could be a link to a virtual reality helmet so that you could get right into problems.

Communications After AD2000. Edited by D.E.N. Davies, C. Hilsum and A.W. Rudge. Published in 1993 by Chapman & Hall, London, for The Royal Society. ISBN 0 412 49550 3

This paper gives an overview of the optical technology which could help to achieve that vision. The paper also looks at the issues which may hinder its development. In order to put future developments into context the paper initially reviews the recent developments in fibre optics.

14.2 CURRENT PROGRESS

It was only just over ten years ago that the very first practical optical fibre links for telecommunications were installed across Loch Fyne in Scotland (Midwinter, 1981), and on links between London and Birmingham. The technology has moved a long way in that time and is still progressing. Figure 14.1 diagrammatically shows the important trends that have occurred over the past ten years. Singlemode fibre has become the first choice over multimode for telecommunications systems. This was an important step since it allowed full access to fibre's potential bandwidth. Coupled with the use of lasers instead of LEDs it has allowed bit-rates in commercial systems to rise from 140 Mbit s^{-1} to more than 4 Gbit s^{-1}. It is now also possible to multiplex several lasers at different wavelengths onto a single fibre. In the laboratory, systems are working with many tens of wavelengths over the same fibre.

Taking the best from all of these developments gives systems theoretically capable of carrying a million telephone calls simultaneously. As a result of these and other developments the system cost has fallen dramatically and consequently the number of systems in use has multiplied many times.

14.2.1 International

Figure 14.2 shows the decline in the cost of a call from London to New York over the past 60 years. In today's money the price has fallen from £241 in 1927 for a 3 minute call to £1.55, while at the same time the technology has changed from HF radio, to coaxial cable, to satellite, and now to optical cables. This demonstrates the effectiveness of new technology in reducing costs. In the past decade optical fibre submarine system technology has been replacing the earlier coaxial cable technology to provide better reach and performance. Current optical cables can operate at 565 Mbit s^{-1}, twice the speed of the best copper systems, with the equivalent of about 8000 telephone circuits, and commercial systems are available at rates as high as 1.6 Gbit s^{-1}. The fibre system also requires only a third the number of repeaters *en route* to counteract the signals fading with distance. Figure 14.2 shows a halving of the price every year on average. At that rate the cost of a 3 minute call in the year 2010 will be just 10p. As connections get cheaper, usage increases, which brings

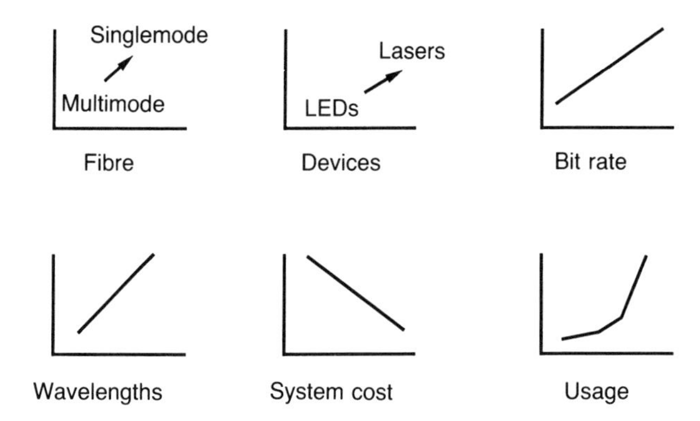

Figure 14.1 Year's progress.

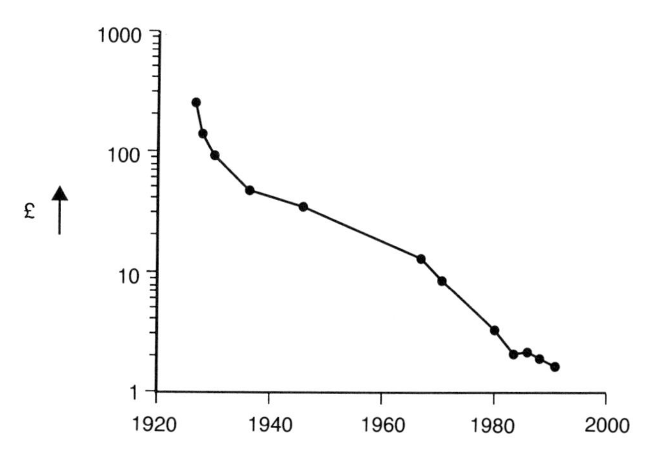

Figure 14.2 Cost of 3-minute call to USA (today's prices).

even greater economies of scale. Perhaps one day transatlantic calls could be as cheap and as commonplace as local calls are today. When that time comes, location will be totally unimportant; people will live where it suits them and work from home or local communication centres.

14.2.2 Business

A fairly recent development has been the use of optical fibre systems to provide a multiplex of many hundreds of circuits to large customer sites (Oakley *et al.*, 1991). This is far cheaper than using the old copper network. It is also far more reliable, and because the bandwidth of the fibre is so great it can easily be expanded as the customers' needs grow;

gone are the days when customers had to wait for new cables to be installed before they could be given new services. Currently there are 70 000 km of fibre installed between major business sites in London. This is the most intensive application of fibre anywhere in the world.

14.2.3 Residential

Fibre is also beginning to be used for residential customers. Currently homes are served by copper pairs which can provide one, perhaps two, telephone connections. Copper pairs have a limited bandwidth and certainly cannot carry broadband signals such as cable TV. Optical fibre can do this with ease. As a result trials of fibre to the home have been set up around the world. One of the most comprehensive is in the UK in Bishop's Stortford (Hoppitt and Rawson, 1991).

This trial has been set up, by BT in collaboration with British industry, to assess the operational and commercial feasibility of cabling fibre directly to homes and businesses. The system provides a range of services including telephone, text transmissions, and multi-channel TV. A key feature of the trial has been the use of passive splitters to enable the cost of expensive exchange equipment and cables to be shared between several customers as shown in Figure 14.3. These devices are potentially very cheap when made in mass volumes, and are perhaps the only way in which fibre can be connected cost effectively to small businesses and to residential customers. The trial has proven the technical feasibility of fibre to the home, but unfortunately economic feasibility is still some way away, and must wait until economies of scale are available. Economies of scale are crucial to the success of a mass market for fibre systems.

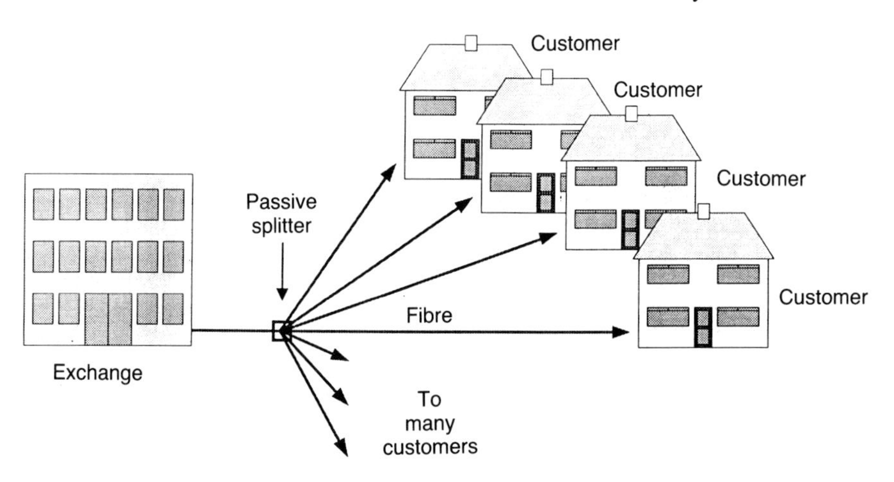

Figure 14.3 Bishop's Stortford fibre trial.

14.2.4 Relative progress

Optical technology is now the norm for international, trunk and for connections to major customer sites. Also, trials of fibre to the ordinary residential customer are taking place. Currently there are 8 million km of fibre produced every year, of which about 45% is for use in North America and 30% for use in Europe. The UK is the largest user of fibre in Europe and installs half a million km of fibre every year. Fibre production is increasing at 15% per year. All this has happened in just ten years and the final stage of getting fibre to the home is inevitable; the only uncertainty is when it will happen.

Figure 14.4 shows how confidence in an idea varies over its life. At the outset confidence grows quickly but soon over-confidence takes over where expectation exceed the idea's capabilities. This is just hype. Once it is realized that the original ideals cannot be met for a reasonable price or in a reasonable timeframe confidence falls as the cynics are listened to. There is then a move towards realism. Having arrived at realism, optimism can then progressively grow as real applications for the idea emerge. Eventually the product is used widely and confidence can be genuinely high.

The position of fibre in various applications is also shown in Figure 14.4. Undersea links are well into universal application, and indeed the old copper systems are now being taken out of service because the fibre systems are more reliable and can carry more calls. Fibre to major businesses is in the optimism region. After a major thrust by BT to install large amounts of fibre in the city of London during a period of hype,

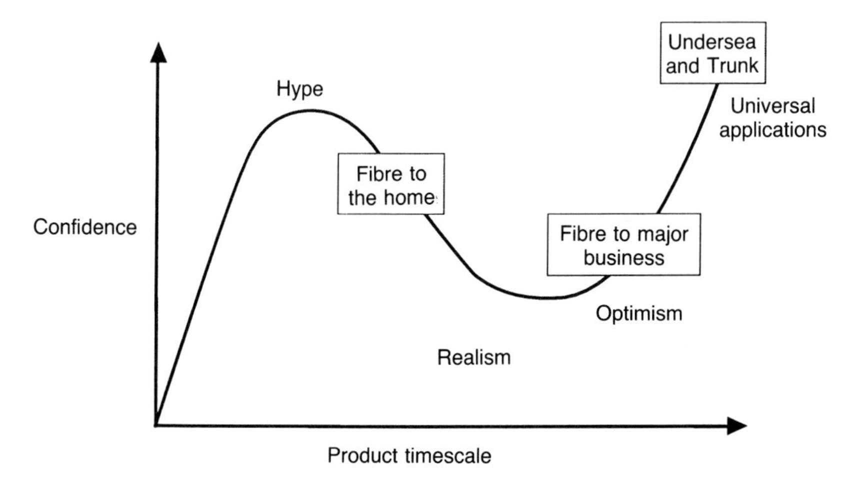

Figure 14.4 The confidence cycle.

it was found that the fibre system were more expensive than original estimates, partly because major changes were needed to the support systems in order to integrate them properly. After a short period of realism there is now optimism that the latest system designs will start the smooth transition towards the universal application of fibre for major businesses. On the left of Figure 14.4 fibre to the home has only just passed the hype region. A few years ago it was widely believed that the nation was on the verge of connecting fibre to every home. However since that time the regulator has prevented BT from freely entering the cable TV market; providing fibre just for a telephone market is uneconomic. The hype is over and fibre to the home is on the downward slope towards realism. Only after the year 2000 will international, trunk, access and residential all be in the true area of universal applications. However even then it will not happen unless the market is correctly assessed, and unless we plan correctly to introduce the right services at the right time at the right price.

14.3 SERVICES

Although technology is a key enabling factor, it should be remembered that it is service which provides the drive. Table 14.1 shows some of the services that are predicted for the year 2000. It is beyond the scope of this

Table 14.1 Services staircase

Today	*1995*	*2000+*
Basic POTS	Wide area centrex	Voice recognition
Linkline	Virtual private network	Video messaging
Callstream	Auto call distributors	Switched broadband
Slow packet	I-series ISDN services	Teleworking
Kilostream	Voice messaging	Business videophones
Megastream	Data messaging	Broadband imaging
Telex	Fast packet 2 Mbit/s	HDTV
Facsimile	LAN-LAN interconnect	Interactive video
E-mail	Global data networking	Go anywhere mobile
Videotex	High resolution Fax	Mobile VANS
Videoconferencing	Videophones for executives	Personal numbers
Slow scan TV	High resolution imaging	User friendly multi-media calls
Cable TV	Improved quality TV (MAC)	
DBS	Telepoint pseudo-incoming	
Cellular	Pan-European cellular	
Outgoing-only Telepoint	Cordless local roaming	
	Data on mobile	

paper to consider them in detail but it worth picking the videophone as a simple example of a dilemma. A high quality videophone has been about to arrive for the past 20 years, but it has been held up by the chicken and egg situation. With no broadband network who will make terminals, and with no terminals who will supply the network?

This highlights the need for integrated planning of new services. A broadband infrastructure will be essential for any future developed country, and indeed in the future the definition of developed country may depend more upon its communications infrastructure than any other factor. Such a broadband infrastructure is essential for the UK, and it is therefore strategically important for BT, the major telecomms operator in the UK. In a competitive environment BT wishes to be in a position to supply new services on demand to its customers. Fibre, and the revenues that can flow from the new services, are essential elements in the strategy.

14.4 TECHNOLOGY

The top part of Figure 14.5 shows today's access network in block diagram form. Copper cables are used from the customer to the exchange, and the switching within the exchanges involves electronics. Copper and electronics are used today because they are cheaper then optical systems, but this limits the end to end performance to a few telephone connections per customer. The lower part of Figure 14.5 shows the bandwidth available at different parts of the link. The optical transmission links between exchanges have a very high bandwidth, but

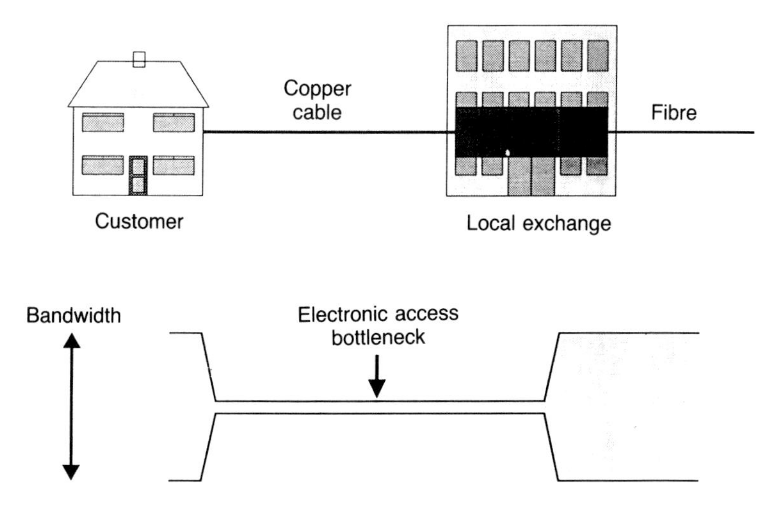

Figure 14.5 Access causes bottleneck.

the rest is limited by the electronic bottlenecks. Ideally it should be the customer who decides what services are needed, and therefore how much bandwidth is required, and not the access technology. Optical fibre, in conjunction with techniques such as asynchronous transfer mode (ATM) or variable bandwidth transmission, can achieve this. Technology trials are under way around the world but until there is a mass market, costs will be too high for anything other than major business use. This again is evidence of a vicious circle that must be broken. The only way to make equipment and services available at the right price is to have the whole industry, and governments, aiming for the same target of a broadband infrastructure for the UK, Europe and the world.

Connecting individual fibres all the way from every customer to the exchange is unnecessarily expensive. The Bishop's Stortford fibre trial showed how parts of the network could be shared to reduce the overall cost. The technique can be extended by cascading systems using recent developments in optical amplifiers to make up for the loss of the splitters. At Bishop's Stortford a 32-way split was used, one fibre from the exchange being fanned out to serve 32 customers. By cascading such networks and using optical amplifiers to make up for the losses it is possible to serve more and more customers from a central point.

Figure 14.6 shows this in diagrammatic form where the system fans out progressively to 32 nodes, then 1024, then 32768, and so on. Laboratory trials (Forester *et al.*, 1991) have already simulated the delivery of 300 uncompressed digital TV channels to over 40 million customers within a 500 km radius, from a single head-end. The experiment demonstrated the possibility of serving half of the country from a single point. Such a system should be able to offer both new and minority services because the initial costs of adding a new service are shared among millions of customers. However, deployment of such a system depends upon the revenue from a mix of services including telephony as well as cable TV. At present the Regulator does not allow BT to offer such a national service and this is a barrier to such new technology being implemented.

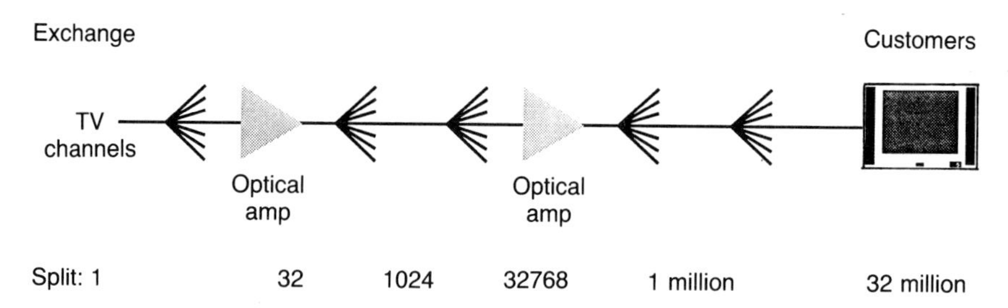

Figure 14.6 Multi-channel broadcast network.

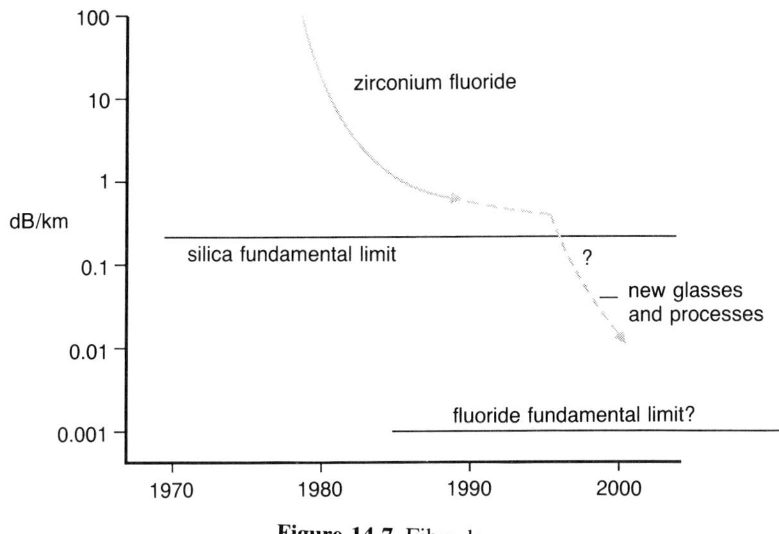

Figure 14.7 Fibre loss.

Figure 14.7 shows how the loss of fibres has come down over time, and it also shows the fundamental limits for different types of fibre. Current optical fibre is made of a very pure glass manufactured from silica sand, and it has a loss of about 0.2 dB per km. Signals can travel about 100 km before they need amplifying. It is interesting to note that this loss was achieved in the early 1980s and has not yet been significantly bettered. Different glasses, in particular fluoride glasses (Day *et al.*, 1990), are now being developed which potentially have a loss as low as a fifth of that of silica fibre. Future systems may have even lower fundamental losses. Using these glasses repeaters would not be needed on a link of 500 km or even more.

However, these new fibres are not yet fully researched; currently the best fluoride fibre has a loss of more than the best silica fibre. The limitations are largely imposed by impurities within the glass. It is estimated that the level of impurities must be reduced to one part per 1000 million. This is genuinely like trying to remove a needle from a haystack – except that the needle is straw-coloured and non-magnetic. If such glasses could be developed the number of repeaters needed for a system to cross the Atlantic would be reduced from about 60 to 12. This would give large increases in reliability and further reduce costs.

In order to get the best out of an optical transmission system the receiver should be as sensitive as possible. The fundamental limit here is the unit of the photon. The ideal receiver would need only a single photon to generate a change in electrical current in the receiver corre-

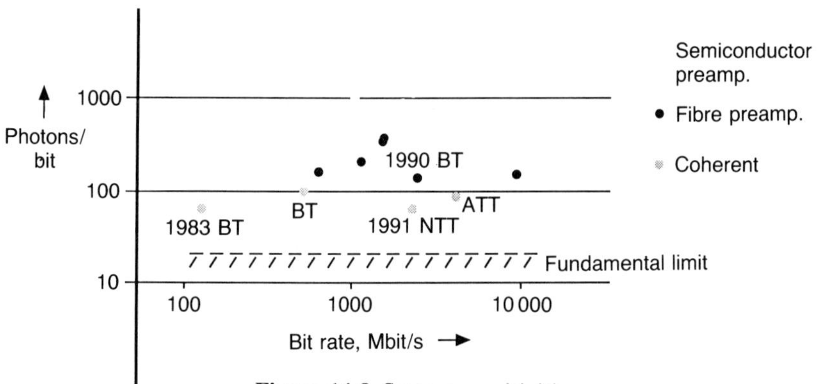

Figure 14.8 System sensitivities.

sponding to, say, a change from a 'nought' to a 'one'. In practice electrical and optical noise cannot be eliminated and the detection process itself is statistical. Although sensitivity can be improved by the use of optical amplifiers and by the use of coherent detection, it turns out that the theoretical limit is about 20 photons per bit. Practical systems using coherent detection have achieved around 66 photons per bit, as shown in Figure 14.8. It is interesting to note that once again this was achieved in the early 1980s and although bit rates have increased there have been few improvements in sensitivity. A recent development has been the use of optical amplifiers (Rowbotham, 1987) as preamplifiers in direct detection systems; these are far simpler than coherent detectors. These systems (Smyth *et al.*, 1990) currently achieve 150 photons per bit and may eventually reach 40 to 60 photons per bit. Their simplicity outweighs the disadvantage of slightly reduced sensitivity compared with coherent systems.

Research work on improving sensitivities is now reducing as the law of diminishing returns takes over. Researchers' efforts are now moving away from lower loss fibre, and more sensitive receivers, towards active fibre which behaves as a continuous fibre amplifier with zero loss.

In order to remove the electronic bottleneck in the network there is also a need to implement optical switching. Optical switching uses the non-linear properties of devices such as lithium niobate and indium phosphide. It is interesting to see the parallel with the transistor. For years all effort was on making transistor circuits with better and better linearity to reduce signal distortion and noise. Now most electronic circuits are digital and rely on semiconductor materials being highly non-linear. The same is true for optics; until recently all efforts were on increasing linearity, but now both optical amplifiers and optical switching need to exploit the non-linearities of fibre and optical devices.

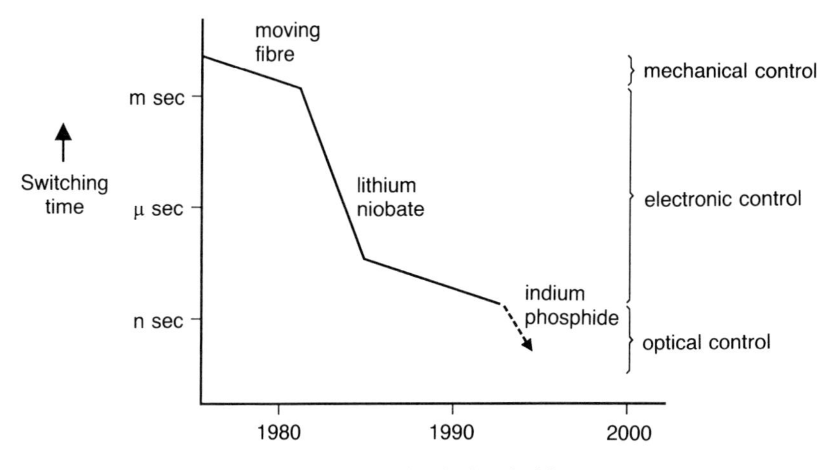

Figure 14.9 Optical switching.

Figure 14.9 shows how optical switching technology has changed with time. Early attempts at optical switching involved mechanically moving fibres or mirrors. These were slow and unreliable. Currently lithium niobate devices are available. These are solid state, but because they are controlled by electrical signals, their switching speed, although very high at around a billion times a second, is still a bottleneck compared to the switching speed which can be achieved by means of pure optical switches. Optically controlled switches using indium phosphide have been developed in the laboratories. One beam of light shines through the device. When another beam is shone onto the substrate, the non-linear change in refractive index causes the path of the original beam to change. Careful arrangement can cause the beam to turn on and off at speeds of around a thousand billion times a second. In the future switching and processing speeds will not be a limitation. With reference to the desk in the introduction, high speed optical processing is needed to switch the huge bandwidths that will be used to serve a mass market of HDTV videophones. In addition, optical processing speeds may be needed to achieve artificial intelligence fast enough to do effective real-time translation and to create realistic images in virtual reality.

A fairly recent discovery related to non-linearity is the soliton. This is a special type of optical pulse which can travel without dispersion. Using solitons it might be possible to send pulses of light through 8000 km of optical fibre across the Atlantic without electronic repeaters or optical amplifiers. Solitons are very complex but the way in which a pulse can travel without dispersion is best understood by thinking of a team of men running on a thick mattress as shown in Figure 14.10. The man at the

Figure 14.10 Solitons.

front is slowed down by having to run up the slope of the mattress, while the man at the back is speeded up by running down the slope. In this way the runners, like the soliton pulse, keep together, and do not spread out, and the pulse travels without distortion. It all depends upon the non-linearities of the mattress and the fibre. This means using heavy men or very high power optical pulses.

14.5 CONCLUSION

Beyond 2000 the technology to deliver almost infinite bandwidth to everyone will exist, and optics will be able to provide almost infinite storage and process power. This should allow unlimited services. Copper and electronics have served telecommunications well over many decades but they can no longer provide the infrastructure necessary for potential future services. The introduction in this paper described one particular vision of the future, and undoubtedly the technology will exist to deliver that vision. However, just because technology enables it to be done does not mean to say it will be done. Deployment of such systems depends predominantly upon cost effective implementations, which require economies of scale. Until there are clear world markets industry is unlikely to invest in the plant to produce the new devices and equipment. Accordingly BT is working hard at an international level to secure such standards as soon as possible.

It is not technology that is the limit. The limit is imposed by capital investment and operating costs. Figure 14.11 shows this diagrammatically. If we are not careful capital investment and operating costs will squeeze the opportunities for new services to a trickle. Although everything is possible we have to open the tap by identifying the market needs using our imagination, and develop the right forward planning capability to get the right services in place at the right time. Government has a corre-

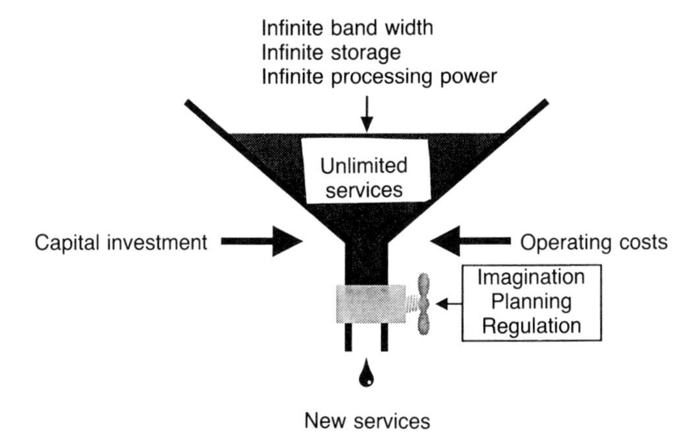

Figure 14.11 Effect of capital and operating costs.

sponding duty to ensure that the right regulatory policy will speed the process rather than hinder it. Strategically there is no doubt that fibre is the right approach. A broadband fibre infrastructure is essential for the future well-being of the UK and its industry, and it is the duty of industry and govenment to do everything possible to open that tap and not to allow it to close.

ACKNOWLEDGEMENTS

Acknowledgement is given to Cliff Hoppitt of BT's Network Planning Automation Section and many other colleagues at BT Labs for their assistance in the preparation of this paper.

REFERENCES

Day, C.R. *et al.* (1990) Fluoride fibres for optical transmission, *Optical and Quantum Electronics*, **22**, 259–77.

Forester, D.F., Hill, A.M., Lobbert, R.A., Wyatt, R. and Carter, F.F. (1991) 39.81 Gbit/s 43.8 million way WDM broadcast network with 527 km range, *Electronic Letters*, **27**, (22), 2051–2.

Hoppitt, C.E. and Rawson, J.W.D. (1991) The United Kingdom trial of fibre in the loop, *British Telecommunications Engineering*, **10**, pt. 1, 45–58, April.

Midwinter, J.E. (1981) Studies of monomode long wavelength fiber systems at the British Telecom Research Laboratories, *I.E.E.E. J. Quantum Electronics*, **QE-17**, (6), June.

Oakley, K.A., Guyon, R. and Stern, J.R. (1991) Fibre in the access network, *British Telecommunications Engineering* **10**, April, 40–7.

Rowbotham, T.R. (1987) The application of laser amplifiers to undersea lightwave systems, *Globecom 87*, Tokyo, Japan, 15–18 Nov.

Smyth, P.P. *et al.* (1990) 152 photons per bit detection at 622 Mbit/s to 2.5 Gbit/s using an erbium fibre pre-amplifier, *Electronics Letters*, 13 September, **26**, (19), 1604–5.

15

Why software engineering at a communications meeting?

C.A.R. Hoare FRS, *Oxford University*

Why software engineering at a communications meeting? This is because it's all really made out of software.

Exaggerating slightly the points made by other contributors to this discussion, one could say that all that vast network of hardware – handsets, displays, wires, fibres, switches, links, antennae, satellites – everything except the physical holes in the ground and the towers in the air – is controlled by software or soon will be; they are just peripheral equipment, as it were, to the computers which run the programs. Software is the magic ingredient which realizes the growing potential of recent and predicted advances in the hardware. It is software that adds value, that assembles the components into saleable systems, products and services. Hardware components will be manufactured in increasing volumes and supplied at reducing cost to all the players in the communications marketplace opened up by deregulation. It is the software that will determine competitive advantage, and distinguish the winners from the losers.

So what is this discipline of software engineering? How does it compare and differ from other engineering disciplines? And how far can it be regarded as mature? These are the questions addressed in this paper; and it will end with an appeal to the communications industries and to higher education to collaborate even more closely in seeking the answers.

The most strikingly visible difference between software and other engineering products is the almost total invisibility of software. There is absolutely nothing photogenic about software, and absolutely no joy in building scale models of its operation. And even less can it be touched, felt, heard, smelled or tasted. It seems to be nothing more than the abstract disembodiment of pure complexity. As a corollary, software is a product where the cost and time required for manufacture and distribution are close to absolute zero. All the expense and delay is in design and

Communications After AD*2000*. Edited by D.E.N. Davies, C. Hilsum and A.W. Rudge. Published in 1993 by Chapman & Hall, London, for The Royal Society. ISBN 0 412 49550 3

development – and later in marketing and sales. Historically, this has made software a very difficult area in which to gain recognition for sound research in universities, or in which to exercise sound planning, management, and control in industry.

But of course the cruder parameters of software can readily be measured. Over the last ten years, such measurements often show that the length of computer programs embodied in a typical product have grown, perhaps by a factor of ten; that they have cost ten times as much to develop, that they are proportionately more likely to contain errors detected in service, and each error could potentially be ten times more damaging in its effect. Hardware has in the same time made equally rapid progress but fortunately in the opposite direction. It is now ten times smaller, faster, cheaper, and more reliable. I really shouldn't have embarked on this comparison, so dreadfully unfavourable and unfair to software engineers.

Let's change the subject quickly, and concentrate on the much more important similarities between software and other branches of engineering. Firstly we share the same goals; they were nicely defined for civil engineering by Thomas Tredgold in 1828 – 'the art of directing the great sources of power in Nature for the use and convenience of man'. Secondly, the success of any engineering project requires full attention to the implications of marketing, commerce, accountancy, management, and even politics. But the single feature that differentiates all of engineering and science from all these other important practical concerns is the explicit, crucial and pervasive use of the techniques and notations of mathematics. Each branch of science seeks mathematical theories, or models of selected aspects of physical reality. The scientist uses mathematics to predict from the theory its observable consequences, which are then checked by careful experiment. The engineer on the other hand uses a validated scientific theory to check the performance parameters of a design before it is put into production. That for example, is why our buildings and bridges no longer blow down very often.

In a mature engineering discipline the direction of the mathematical calculations is reversed. Start with a mathematical model of the customer's requirements. Decide the general strategy and structure of the solution; and then with the aid of calculation derive the content and detail of the design, including optimization of relevant parameters. The design now needs no further test or check – it is correct by construction. This reversal from bottom-up predictive mathematics to top-down design calculations is the goal of all research into engineering method, in all branches of engineering, and especially software engineering.

But this simple story has ignored the incredible complexity of the symbolic and numerical calculations required by modern science and engineering. To limit this complexity, science presents not just a single model of reality, but rather a whole hierarchy of models, dealing with phenomena at differing scales, at differing degrees of granularity, and at differing levels of abstraction. For example, starting at the level of chemistry, physics offers the hierarchy

- molecular dynamics
- atomic theory
- elementary particles
- chromodynamics

Each level has its own autonomous concepts and its own model, which can be understood and used independently of the others. It is the general goal of the pure scientist to secure the links between the levels, defining the concepts at each level in terms of the previous one, and then proving its axioms as theorems in the previous theory.

A similar hierarchy of levels of abstraction is equally necessary in engineering. For example in design of computer hardware we separate

- instruction set
- register path
- microcode and control
- combinational design
- switch level design
- circuit electronics

Here again, at each level there is a different conceptual framework, a different notation, and a different calculus of design. A complete design at each level of abstraction serves as a specification for the design at the next lower level. It is the particular goal of the practising engineer to ensure that a specification at each level is correctly and efficiently implemented by the selected design at the next lower level.

Communications engineers engaged in the design of protocols are familiar with the famous seven levels of the international standard

- application
- presentation
- session
- network
- transport
- data link
- physical level

The general principle of the hierarchy has been very successful. At the higher levels, there has been some delay in finding and agreeing on the appropriate abstract concepts for formulating the standard; but this kind of conceptual engineering is the necessary condition for breakthrough in any branch of science, or indeed any kind of intellectual endeavour. Just saying that is never going to make the discovery easy to make, however simple it may seem afterwards. At the lower levels of the protocol hierarchy, maintenance of the design structure throughout the implementation can cause problems of efficiency; but these are solved with the aid of correctness-preserving transformations, which combine the benefits of structured specification and design with highly optimized implementation. Software engineers can learn a lot from this transformational approach to specification and design.

Software engineers also have a hierarchy based on scale and granularity. They talk of

- systems
- modules
- classes
- objects or processes
- functions and subroutines
- straight line code
- individual instructions

The mathematical theories which are useful for design calculations at each level have been to a large extent developed by software engineering research, and the transitions at the lower levels have been successfully formalized and even automated. The results of this research are being gradually assimilated into industrial practice. Progress is slowed by a general antipathy to mathematics among software engineers – but this feeling is yet another characteristic which is shared with other branches of engineering, and indeed with most of management, and with the general population.

Mathematics itself provides an outstanding example of the control of complexity by structure and abstraction. Its branches can be arranged in a hierarchy like those of physics – topology abstracts from analysis; analysis provides the basis for calculus; and calculus can be used by engineers who understand nothing of the more abstract foundations. Even within a single branch of mathematics, a lemma can be safely used without studying the complexity of its proof, a theorem abstracts from the complexity of its lemmas, and a theory from the collection of its separate theorems.

Similar structures are observable in good programming practice, where larger programs use smaller ones as subroutines, or subobjects, or subprocesses. But this analogy is good only if the formal statement of the function and purpose of each subroutine is as simple and complete as that of a mathematical theorem, and an order of magnitude simpler than the code which implements it. Furthermore, the reliability of the code must also approach that of a mathematical theorem. A large system constructed from even slightly unreliable components can rapidly collapse, either before or after delivery. Reliability is the very essence of engineering, and it is achieved by explicit appeal to the concepts, methods, abstractions and structures of mathematics.

But, ironically, it is not to the traditional applied branches of continuous mathematics that the software engineer turns for guidance, but rather to the traditionally pure branches of discrete mathematics – set theory, algebra, and even category theory. The reason is that software engineering deals very largely with discrete phenomena, transitions, events, values and structures. At the lowest level we have just the two discrete boolean values, zero and one. In a large program if just one of these digits is changed, even only for just one millisecond, the consequences for the whole program, indeed the whole system, are in practice quite unpredictable, and in principle potentially disastrous.

This means that the software engineer cannot rely on smoothness or continuity in the control of tolerances or error. Numerical approximation is simply not available as a technique to simplify calculations. Since there is no appropriate metric, worst case analysis and worst case testing are just not available. For the same reason there is no freedom to get it nearly right; even if there were, there is no way that this would simplify the task of design and implementation. Approximation, even to the extent of order-of-magnitude calculation, is the stock-in-trade of the engineer, the most important way of maintaining intellectual control at the early stages of design and throughout the later implementation. In the discrete branches of engineering, for these purposes we have to rely almost wholly on structure and abstraction.

The gap between continuous and discrete engineering is one that puts nearly all modern telecommunications and electronic hardware design on the same side as software engineering – certainly all of network design down to the individual switch, and all of VLSI design, down to the individual logic gate. To cover this range of disciplines perhaps I should use a more neutral term like discrete systems engineering, or a more fashionable one like information engineering. Under whatever title, I believe that we will see by the year 2000, a strong convergence in the practice of engineering of software, hardware and communications. The communications industries will be the first to realize and benefit by this convergence.

This convergence of communications, hardware, and software engineering is simply and elegantly illustrated in mathematical theories known as process algebras, developed over the last twenty years by basic research in universities. The theory combines the concepts of conventional sequential programming with the kind of concurrency which is embodied automatically in every combination of hardware components, and the kind of communication which occurs almost naturally whenever hardware components are connected by wires. The theory has already served as the basis of a draft international standard (LOTOS) for the definition of protocols, for the design of a programming language (OCCAM), and a microprocessor (the transputer), and the design of several silicon compilation languages. But the mathematical theory is much more general; with slight variations, it can be applied on every scale and at every level of abstraction and every level of granularity, from the customer's view of the services required of the system as a whole, through the design of the major components of the network and the interfaces between them, down to their implementation on a collection of processors and special purpose application specific hardware, interacting with each other at any distance. It is this appeal to abstraction that permits theories tested in the laboratory to be cautiously scaled up for industrial application. The uniformity of the mathematical foundation permits all stages of the design and implementation to be related by calculation and proof. Many of the stages of adaptation and optimization can be codified and carried out (or at least checked) by automatic transformation systems. It is this that promises not only an increase in the quality and reliability of the product, but also a reduction in the time to market.

It is experience with this kind of practically applicable theory that leads me to predict that discrete systems engineering is making rapid progress towards the status of a mature engineering discipline and maybe will reach it by the year 2000. A fully mature discipline will have the following characteristics:

- it puts the customer first;
- it codifies corporate strategy;
- it puts management in control;
- it magnifies human intellect;
- it builds its own tools;
- it is the language for professionals; and
- it is transmitted by education.

A mature engineering discipline puts the customer first. It starts with a scientific investigation of the actual characteristics and behaviour of the customer population, not just as individuals but as members of their societies; in the workplace, school or home. It takes full advantage of the methods and results of the human sciences – physiology, psychology, linguistics, sociology. It analyses stated requirements and stated assumptions until a clear picture emerges of some desirable future product or development that will satisfy the true requirements, which have often been left unstated. The mismatch between perceived and actual requirement is one which must be overcome by good marketing. Meanwhile, the engineer must attempt, as soon as and as far as possible, to construct a faithful mathematical model of both the numerical and discrete properties of the projected and desired interactions between the community of customers and the projected product or service. This is the first and most important interface to define; it is the basis of all subsequent engineering design, and any lapse of judgment here could lead to a product that is undeliverable or unusable. The pace of change is no longer driven solely by technology: in software especially, the technology must be driven by the customer.

A mature engineering discipline formulates strategic policy. No large enterprise can afford to design and deliver a single product at a time, no matter how advanced the technology or how timely its introduction to the market. The real challenge is to design an architecture for a family of products, covering not one but a range of markets, with not just one product in each market but a series of complementary, supplementary, enhanced and eventually replacement products, stretching into the foreseeable future. The strategy must be presentable in abstract terms at board level, so that it can be correlated with financial management, marketing, resource planning, and ultimately with the image that the enterprise wishes to have of itself. The engineering discipline must provide the appropriate abstractions and theories to define the structure and interfaces of the entire family long before any of the detailed design begins; and this must be backed by enterprise-wide engineering standards which give assurance that the strategy can be implemented as planned. The days of innovation as adventure are over. In the framework of strategic policy, innovation is routine.

A mature engineering puts management in control. Each level and branch of management can understand, within a self-contained intellectual framework, all the objectives and activities of subordinate levels, and can take confident responsibility for the way in which these contribute to the goals of superior levels. The confidence is based upon abstract but precise formalization both of the vertical and of the horizontal interfaces throughout the management hierarchy. The confidence is justified by mathematical calculations, which establish in advance of implementation that if each subordinate goal is met, the superior goal is guaranteed. Complexity is controlled by correlating levels of management with levels of abstraction, so that problems and delays can no longer be hidden under a morass of technical detail. In spite of the intangibility of software, signs of trouble are immediately visible, and if change is required in the design or in the interfaces, the manager can explore and report all the wider consequences of the change before authorizing it. As a result, there are few last-minute surprises. When the components are delivered, they slot together without prolonged integration testing, they can be delivered immediately with minimal risk of feature interactions discovered in service. 'Design right first time' is no longer a slogan but has become a habit.

A mature engineering discipline releases the full potential of the human intellect. Because specifications are expressed at the highest possible level of abstraction, they give the widest possible scope for exercise in design skill, ingenuity and inventiveness of the human engineer. The mathematical theory defines the boundaries of the design space, and provides the method by which it may be thoroughly explored. As new ideas emerge, they can be crystallized with the aid of mathematical formulae, which can be objectively discussed, evaluated, justified or even admired. Finally when the design is frozen, it can be made sufficiently precise for reliable implementation by teams of less experienced or inventive engineers or technicians.

A mature engineering discipline constructs its own design tools. The validity of the methods which transform specification to design and design to implementation is assured by their basis on the well-established scientific theories which underlie the discipline. Only parts of the tool are fully automatic; at all crucial stages, guidance is needed from the skilled and experienced engineer, who has the understanding and inventive talents to direct the design towards a cost-effective solution. The only contribution of the tool will be to calculate of few parameters, and to organize the mass of associated detail in a manner which ensures correctness by construction. In future, the programming of individual lines of C-code will seem as archaic as laying out individual transistors and wires on a silicon chip. But even when a tool is really successful, the general impression should be that it only does the easy bits.

A mature engineering discipline provides a language for communication among professionals specializing in its various branches. The underlying mathematical theory not only explains their common foundations, but explains why and for what purposes it is necessary to differentiate them. There is no longer any need for clamorous conflict between the various branches of software engineering, each claiming exclusive merit for a single computational paradigm: the functional programmers, the logic programmers, the object-oriented programmers, and the no-nonsense hewers of hardware or hackers of C. Any large system will have components constructed from a variety of technologies; and the interfaces between the technologies, which is where most of the problems of engineering arise, are controlled by the abstractions of the underlying general theory.

Finally, a mature engineering discipline is transmitted to future generations of engineers by further education. Its theoretical foundations, by their abstraction and elegance, can be taught as a free-standing mathematical discipline at university or even at school. Its methods and principles can be illustrated on a small scale by student demonstrations, experiments and exercises. By repeated exposure at many different levels to the transition between abstraction of specification and details of implementation, the student comes to understand how the techniques generalize to an industrial scale of application. When this education has been complemented by a period of industrial experience, the educated engineer is intellectually equipped to rise through the management hierarchy to the very highest levels.

This brief and idealized account of engineering education contrasts strongly with the training on the job, which was the only training available when I entered the profession in 1960, and is still the norm today. We learnt programming in a wholly operational fashion, by trying to understand the behaviour of the computer which is executing the program. Execution traces are the only means we had of understanding and removing errors. Errors were regarded as inevitable, because we had no technology to avoid them, even in principle. We hardly recognized the possibility that a complex program might have a simple specification, of far greater benefit to the customer than the implementor. Lengthy and total immersion in operational detail actually inhibited progress towards understanding of the necessary simplifying abstractions.

When it became necessary to learn a new programming language or use a new operating system, training was based on the voluminous manuals which accompany the software. Because there was no common culture or education in the understanding of abstraction, these manuals too have to be based on the lowest level of operational detail. Their volume, complexity, and structural deficiencies absorb all the intellectual energy of the student; and yet they were so incomplete or even inconsistent in detail that, when used in earnest, the only way of finding out what the software will actually do is by experimental trial and error. The tool which should be helping has become part of the problem.

The absence or even conscious avoidance of mathematical abstraction in programming education explains why many programmers have often been regarded more like craftsmen or technicians than engineers. They are wonderful people, with experience and skills greatly to be admired and valued. But they work best in isolation on self-contained tasks. They have no language to discuss, explain and justify their work to their colleagues and superiors. Documentation is their bane. They do not read the technical literature to keep abreast of their field. On promotion, they find it difficult to establish or maintain intellectual control of the work of their teams. That is why it is rare for the best programmers to rise to the higher levels of management. Yet it is not conducive to the health of the enterprise when the worser ones do.

The transition between a craft and a mature engineering discipline is always fraught with confusion, difficulty, animosity and charlatanism; and the intangibility of software has certainly prolonged the agony. Most of industry and commerce in Britain is prepared to wait for magic solutions to emerge, and then to buy them in from across the Atlantic. But the more far-sighted enterprises can see the competitive advantage to be obtained by rapid transition to an engineering attitude towards software. Among them, the telecommunications industries, for motives which have been explained elsewhere in this book, are playing a leading role. It is essential to them to raise the educational level of their software engineers, by in-service courses for experienced programmers and their managers, by promoting the quality and relevance of the subject in higher education, by promoting research at the interfaces between technologies; and by attracting the very best of the graduate population into their teams and eventually into their management. Divergence of culture between traditional communications engineers and the new software engineers educated at leading universities must not be allowed to hinder the flow.

This is my appeal to the telecommunications Industry as a whole: and perhaps this is the true reason why I have been invited to contribute to this work. I am very grateful to the organizers for this opportunity to do so.

I am grateful also to Elspeth Cusak, David Freestone, Charles Jackson, and Sinclair Stockman (all of British Telecom) and to Joseph Goguen for assistance in the preparation of this paper.

16

Improving communications at the desktop

Dr A. Hopper, University of Cambridge, and Olivetti Research Ltd

16.1 MULTI-MEDIA COMMUNICATION SYSTEMS

On a typical desktop of today we find a telephone and a computer. A regular telephone is used for audio and the computer provides communications in the form of electronic mail and sometimes facsimile. We may think of the user interface as being an electronic desktop. In future, the telephone will provide additional features and higher quality sound. It will probably still be quite small and with some form of display. The workstation will also be able to handle audio, but additionally video will pass through it. Such multi-media devices are likely to be well connected to others, both in the local area and the rest of the planet. Figure 16.1 shows some of the networks, devices and applications we may expect to find in communication systems which combine various form of media, in particular audio and video (I.E.E.E., 1990; 1991*a*).

Networks

Because audio and video applications are considered important, many networks have properties suitable for multi-media designed in from the outset. It is advantageous if the network provides good jitter control so that synchronization can be moved to the source end. In general, interleaving at the frame level is not adequate with normal frame rates and interleaving has to be done at the sub-frame or tile level. The asynchronous transfer mode (ATM) approach is well suited for this purpose and may become widely used. ATM networks of various kinds are being designed from infra-red systems to electronic systems using high capacity switches.

Communications After AD*2000*. Edited by D.E.N. Davies, C. Hilsum and A.W. Rudge. Published in 1993 by Chapman & Hall, London, for The Royal Society. ISBN 0 412 49550 3

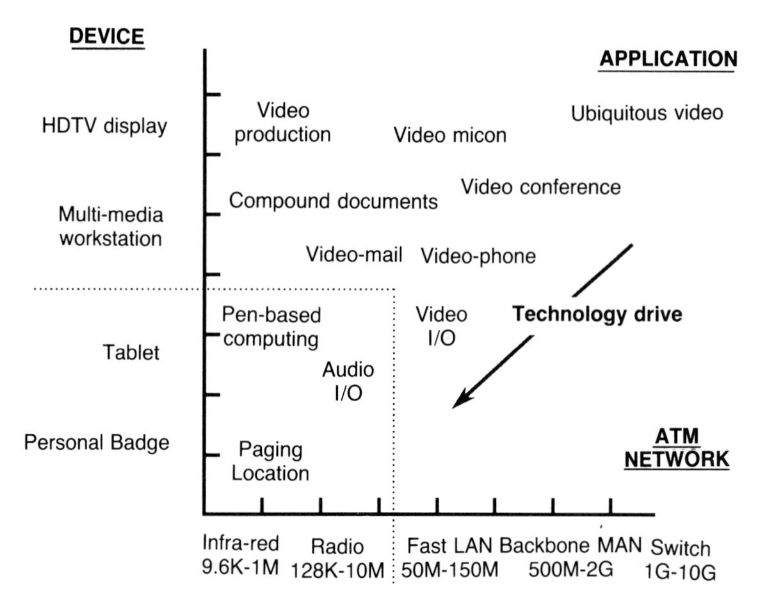

Figure 16.1 The multi-media communication space.

Devices

Devices carried by users are very much affected by their physical size and convenience of use. The smallest may be a personal badge which, used in combination with infra-red or radio systems, makes possible location and paging applications. A display tablet is larger and allows the pen-based model of computing to be used so that communication combining audio with writing on a shared screen may be possible. Until recently, workstations required special hardware to handle audio and video. A threshold is now being crossed where both CPU power and storage capabilities on standard machines are such that vision and hearing can be satisfied using largely non-specialized, and thus volume-produced, hardware. A workstation attached to a fast LAN can provide video-mail and video-phone applications, while a more complex workstation will be able to handle many multiple streams in parallel. A combination of a very high speed network and a HDTV workstation display will enable multiple video streams to be handled in a totally omnipresent way.

Storage

The storage devices provided for multi-media systems will be used both for storage of conventional data and for handling of synchronized streams for real-time applications. The design of such storage systems has to be oriented both towards bursty high speed transfers to one device at a time, and also to well controlled parallel delivery of continuous streams. The digital representation of multi-media signals will mean that files have to be arranged in ways which gives fast random access. The spare CPU power of such file servers will be used to improve the quality of the stored images by software digital signal processing. It may also be used for generation of navigation information and annotation of real-time streams to make subsequent retrieval easier.

Compression

Minimizing the information loss at the transmitter allows the receiver better control of how to interpret, scale or use the data. In some applications, such as facsimile, only point-to-point transmission is envisaged and so a large amount of compression can be achieved for that single transfer. In computer systems arbitrary transfer of data between different points takes place and thus it is crucial that the compression systems should make this possible. In particular, it is necessary to consider the case where a particular stream is passed several times through a cycle of compression, decompression, and manipulation. Information loss has to be minimized or made to approach some asymptote. Compression can also make it difficult to combine different types of data such as video and graphics.

Types of traffic

Because network traffic needs to be real-time only if an individual is viewing it, other transfers can take place at either higher or lower than real-time speeds. If the camera becomes widely used the amount of video and audio data that will be present will greatly increase. This bulk data will be transmitted across networks in conventional ways and may dominate. Another type of traffic will be location information used to indicate the presence or movement of objects. This traffic has its own real-time restrictions because delay in reporting movement of objects may render the information useless.

16.2 DESKTOP CAMERA

A project started in 1987 at the Olivetti Research Laboratory and the University of Cambridge Computer Laboratory has dealt with the design and use of systems for desktop multi-media applications. The objectives of the Pandora project were to investigate how to construct such systems and also to find out if using a camera or microphone on the desktop as part of the normal computer environment is useful.

16.2.1 Pandora architecture

The Pandora Project set out with a number of goals and constraints. It was decided that the system would be completely digital and that the video and audio would be of medium quality. No hard restrictions were to be placed on the number of real-time streams it could handle. All streams were to be networked and switched in unconstrained ways subject to the limit imposed by total system performance. The system was based on a Unix workstation and so the screen for delivery of images was the same as one for the delivery of windows and conventional computer applications (Hopper, 1990; 1991).

In order to implement a flexible system it was necessary to adopt a hardware-oriented approach. This involved the use of a video and audio subsystem called Pandora's Box attached to the Unix workstation. For each pixel Pandora's Box can choose to pass through the original workstation pixel or to introduce a pixel of its own. There is no restriction on the pattern and shape of replaced pixels although in practice only rectangular video windows are used. As well as handling input from a camera, Pandora's Box interfaces the microphone, speaker and network. The hardware itself consists of several transputers, controlling capture, display, audio and network cards. A low level kernel provides the switching control. The kernel has a general purpose interface through which the workstation can request streams to be set up as required. An ATM networking system was adopted with a capacity of 500 Mbps in the backbone with distribution on local sites being done at 50 Mbps (Greaves *et al.*, 1990).

Table 16.1 shows the bandwidth requirements of the more commonly used video and audio formats on the Pandora system. When the streams are transmitted, an overhead of about 5% has to be added for protocol headers. The quality of the images was chosen so that the ATM network could support many such streams without interference. Similarly the workstation display can support many streams without constraint. The bottleneck in the system is the interface between Pandora's Box and the network; this can support about 5 Mbps of real-time traffic in arbitrary combinations. As well as handling video, Pandora's Box provides audio mixing so that up to about five digital audio streams can be received by the box and mixed correctly for output through a single loudspeaker.

Figure 16.2 shows the Pandora System in use at present. About 25 Pandora's Box workstations are dispersed across two sites, half a mile

Table 16.1 Pandora stream properties

	size	*frame rate*	*compression factor*	*required bandwidth*
normal video	256 × 256 × 8	25	8	1.64 Mbps
medium video	128 × 128 × 8	25	4	819 Kbps
small video	64 × 64 × 8	12.5	2	205 Kbps
audio	8 bit μ-law	8K		64 Kbps

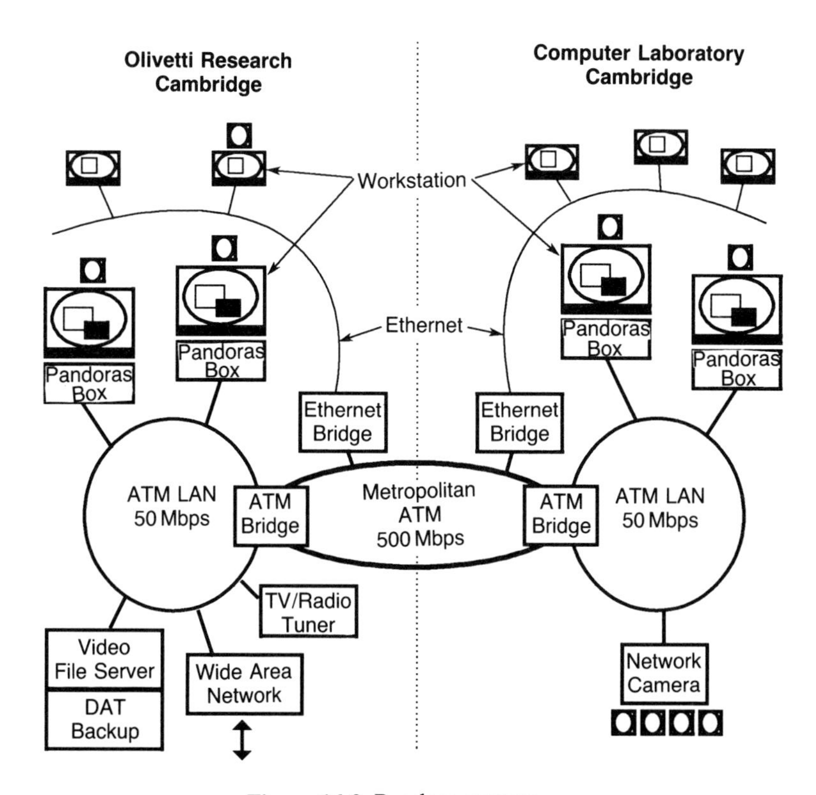

Figure 16.2 Pandora system.

apart. The Pandora System provides services such as TV, multi-media file servers, and bridges to conventional LANs and wide area networks. The system has been in use for about 18 months in a community of about 50 users of which half use it regularly.

Table 16.2 Summary of applications

General points		*Acceptability*
local image not mirrored		high
lip-sync with own image		?
no eye contact		medium
low quality threshold		high
Applications		*Popularity*
Using live streams:	2-way video-phone	high
	3-way, 4-way video-phone	low
	live information services (TV, radio)	low
	permanent video links between sites	medium
	peek in office (bidirectional)	medium
	look everywhere	low
Using storage:	video-mail	very high
	system documentation	low
	general material	medium
	personal introductions	low
	latest news	medium
	mixed text/video mail	very low

16.2.2 Pandora applications

The way the Pandora system and applications have been used is summarized in Table 16.2. With a small community of users it is not possible to extrapolate to larger groups, although it is likely that some of the experience will be repeated elsewhere. However, many of the predictions we made have, in practice, turned out to be inaccurate and the new facilities became popular for quite unexpected reasons.

General

At first users may have to rearrange their office to give good lighting and they will have to become accustomed to a camera. Pandora applications seem more natural in hands-free mode which requires careful consideration of the acoustic properties of an office or a room. The user speaks towards the workstation which has a microphone and a loudspeaker nearby. Feedback control and echo-cancelling circuitry is used to improve sound quality.

In the Pandora System a local image is shown as a true image and not a mirror image. A user of the Pandora System quickly gets used to not treating the screen as a mirror. This is fortunate because a true image has to be transmitted and generating both true and mirrored images in hardware would be expensive. Inevitably there is a slight delay between movement of the lips and the electronic version as presented on the local screen. Networked streams showing other parties can be well synchronized.

Because conventional workstations are used, the cameras are to one side or above the screen rather than being behind it. The resolution is such that lack of eye contact is not immediately apparent. Users are prepared to accept a lower quality threshold than was anticipated. High frame rates with lower resolutions are preferred to better quality images which appear jerky because of low frame rates. Break up of the audio streams renders the applications unusable.

Applications using live streams

The video-phone is the most popular application using live streams. In a video-phone the user sees a local image of himself and uses this to centre his head on the screen. Most users prefer to see a small image of themselves, finding a large one distracting. A large image of the other party is shown in another window and hands-free operation is almost always used. The receiver is invited to join a video-phone conversation by a window appearing on the screen indicating that a call is being requested. There is a choice of several replies: accepting the call, asking for the call to be made again later, or indicating that a return call will be made. This has turned out to be a particularly useful feature and makes a video-phone a more acceptable piece of office equipment.

Two way video-phone conversations are popular and typically last longer than telephone conversations for the same users. At Pandora resolutions the body language is passed over well, and it is possible to tell whether the correspondent is still interested in what is being said. In effect, conversations can last for quite a long time because of this feedback. Figure 16.3 shows the distribution of video-call lengths. Most calls last between 1 and 5 minutes although there is quite a long tail in the distribution and some calls go on for 15 minutes or more. The shape of this graph has largely remained unchanged with use of the system. Three and four way video-calls are less popular, but this may be because the community of users is not large enough.

Another live application is the provision of media information services which combine video and audio. Initially it appeared that these would be popular but in practice they are hardly ever used except in times of crisis or some significant news event.

It is possible to set up a video link between different places and leave it

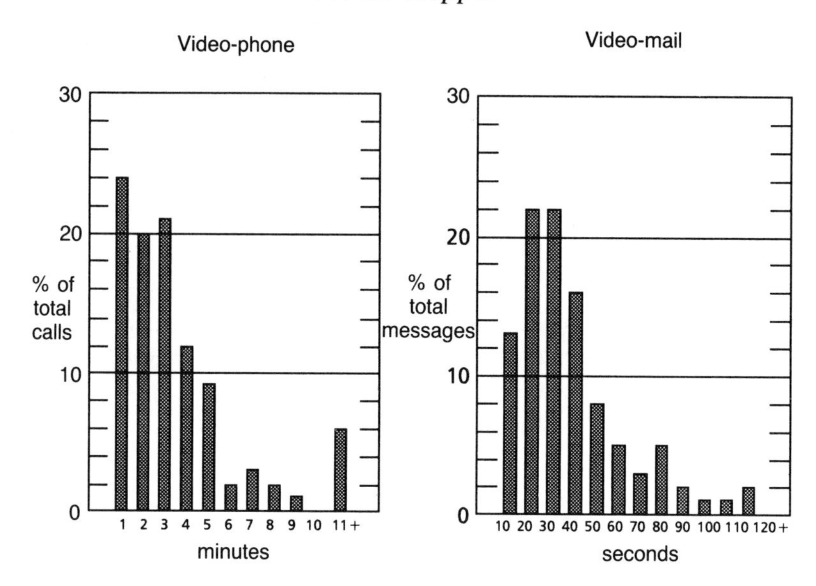

Figure 16.3 Pandora statistics.

running. Normally these are bi-directional and it has been observed that from time to time such links are set up between public areas. The users comment that even though they know where everybody is (see Active Badge section later) it is nevertheless useful to have some feeling for what is going on in another room. When invoked between private offices bi-directional video links can be used to peek into another office. The parties see a picture of each other on the screen. This is usually followed by a smile, a wave and shutdown after 3 or 4 seconds and is moderately popular.

It is possible to receive video streams from all cameras in the system at once. This application is only used for demonstrations and does not seem to fulfil any useful purpose.

Applications using storage

The multi-media file server provides a general purpose storage facility for audio and video. It can support about five simultaneous streams on the network interface. The most popular application using the file server is video-mail, which involves the recording and transmission of a video/audio message. A delivery message is sent to the receiver and the video-mail is viewed when convenient. Most messages begin with a courtesy and then have a substantial part with a nuance or several conditional points which, if written, would require careful drafting. Figure 16.3 shows the distribution of video-mail message lengths. As with video-phones the shape of this graph has not changed substantially in the lifetime of the system. Video-mail longer than about two minutes is considered unduly long although there is no upper limit imposed by the system. It is interesting to note that no video-mail 'flaming' takes place. For a number of users the video-mail system has become the main communications channel although they use text where accuracy is required. Video-mail is quite different to and complements other forms of communication.

The file server currently has 2.5 GBytes of storage which represents about 6.5 hours of typical recordings. There are no restrictions placed on how this facility is used and more storage could be made available if required. About 20% of the contents represent video-mail which has a short lifetime of a few days for each piece. Material which is available to everybody and kept much longer includes systems documentation and general items. Some users like to record personal introductions which contain any message about themselves they wish. Thus while the great majority of recordings that have been made represent video-mail the storage system is mostly filled with other material.

There are a number of applications which combine live streams and storage. One of these is the automatic recording and re-recording of news, weather and other live programmes to the file server. The file containing latest news is updated at regular intervals throughout the day. Thus every user has the latest news available at all times. It is the ability for the user to choose the time when to view that makes this application more popular than the live service.

It has been a surprise that a system which allows users to combine audio, video and text has not proved popular. In this application the normal mail handler is used to generate text into which video clips are inserted as required. The whole compound document is sent to another party who can read the text and view the video parts. There may be a number of reasons why this application is not popular. The first is that the manipulation tools which are presently available are not good enough to make it easy to generate such documents. An alternative view is that video-mail and text-mail provide different methods of communication and it is quite natural to make a choice of the one better suited at a particular time. For the majority of communications the video-mail system is appropriate because it is quicker, but where particular accuracy is required text-mail is used instead.

The Pandora environment enables new multi-media communication applications to be evaluated. For example an evaluation is being made of switching of additional streams during video-phones so that video-mail or other material can be shown in a video-phone session. However, at present re-use of material on the file sever is difficult because there is a lack of retrieval and manipulation techniques for multi-media data.

16.2.3 Extending the use of desktop cameras

Because video-mail does not require a high speed communications path from source to destination it is possible to deploy a system which uses the existing networks for distribution. The recording is made to a local disc (or network server) and from there the data is sent as a normal file transfer. The video-mail can be viewed on standard computer workstations because many have audio output and a fast enough frame store. To generate video-mail on standard workstations a simple capture card for digitizing the camera and microphone output is required. At present experiments are taking place with use of video-mail between research laboratories in Cambridge, (England) and the USA. It is too early to say how this use of video-mail will evolve but early signs are that it will be popular.

When the Pandora system was being designed, the multi-media features were only feasible for fixed workstations. However, it is now becoming possible to design a portable system of similar performance. This would extend use of Pandora applications in the local area. It will be interesting to see what types of communication device users are prepared to carry.

Recently CCD technology has become sufficiently mature to appear in combination with other logic on single chips. This will make it possible to design units with on-board digital processing to provide digital video in very flexible ways. Cameras will become ubiquitous, making feasible applications from simple observation systems to complex data gathering systems. A new project called Medusa aims to extend the use of cameras at the desktop and in a system. The assumption is made that many more cameras will be associated with each user and multiple networked streams will be generated. Thus a workstation might have eight or sixteen cameras, while a video system used in public areas might have many more.

In the Medusa system the receiver will have control over all streams and, subject to the sender's authorization, will be able to transmit them to a recipient for selection of which one to watch. The recipient may be aided by local hardware, software or hints from the transmitter in choosing the required stream. The principle of receiver control is important and applies not only to choices between multiple video and audio streams but also to representations for data of all types. While at present this is technically difficult because the data rates generated may be high, in due course, when the capacity of networks increases, it may not be a problem.

A multi-stream capability of this kind can be used in a number of ways. If the cameras are attached to a workstation directly then each one could point in a slightly different direction so that centring the head can be automatic. Alternatively, the cameras could be arranged to cover a particular volume of space so that the user can move around and think of the cameras as being omnipresent in that space. Another way of using multiple streams is to use each one for different types of information. For example, some of the cameras could be sensitive to the infra-red and may be able tell where hotter parts of the image are, which may be useful for applications such as tracking of faces or other objects. It may be possible that an infra-red grid can be projected and then viewed by a infra-red camera to obtain depth information from the distortions. Yet another way of arranging the cameras would be for use with three-dimensional displays. This would require all the sources to be set up in a coherent way such that the requirements of the 3-D display are met. A more conventional alternative for Medusa would be to use all the capacity to drive a single HDTV display.

16.3 DESKTOP INFORMATION

A way of improving communication at the desktop is to give a better idea of who is available, who is not, and what the user communication preferences and choices are. The Active Badge project aims to investigate the design and use of a small personal device to make more attractive the use of communications (and computer) facilities.

16.3.1 Active badge system

The Active Badge is used to provide information about where people are (Want *et al.*, 1992; Want and Hopper, 1992). It is battery powered, transmits in the infra-red spectrum and is approximately 60 × 60 × 8 millimeters. The transmissions take place every 15 seconds and identify the badge. There is a single button on the badge which causes an immediate transmission. Receivers are linked by wire to a computer and are placed so as to define cells for the coverage required. Normally they correspond to spaces occupied by one or a number of people. The range of the infra-red system is about 20 metres and communication does not need to be line of sight. The badge has a light dependent resistor used to reduce power consumption by decreasing the frequency of transmissions when in the dark. This also means that the user can switch the badge off by placing it in a pocket or face down on the table. Not all badge transmissions are picked up by a receiver, but by using simple algorithms in the receiving software the system can be made sufficiently accurate to be very useful.

Table 16.3 Active badge display

Name	*Telephone*	*Position*	*Seen*	*Status*
P Ainsworth	343	Accounts	Static	Alone
M Chopping	410	R410 MC	Friday	
D Clarke	316	R316 DC	12:30	
D Garnett	218	R435 DG	12:20	
T Glauert	232	R310 TG	35 mins	
S Gotts	0	Reception	Static	With Jackson
D Greaves		Floor 3 Corridor	Moving	Alone
A Hopper	334639	Univ Comp Lab R76	Static	
A Jackson	0	Reception	Moving	With Gotts
A Jones	210	Meeting Room	Static	At a gathering
T King			Yesterday	Away in Italy
J Martin	310	Machine Room	24 Dec	
O Mason	210	Meeting Room	Static	At a gathering
D Milway	BUSY	R211 DM	Static	Alone
J Porter	398	Library	Static	Alone
C Turner		Front Door	Yesterday	
R Want	308	R215 RW	7 mins	
M Wilkes	210	Meeting Room	Static	At a gathering
S Wray	204	R212 SW	Static	Alone

14:09 Tuesday 8 January 1991

Many different applications which use the badge sighting information can be devised. One of the first was the provision of location information about individuals. A typical interface is shown in Table 16.3 and gives the name of the person, his position and the number of the nearest telephone. The column marked 'seen' makes it easy to distinguish between individuals who are static and those who are in the process of moving from one place to another. If the badge has not been sighted for three minutes, this column shows how many minutes ago the badge was last sighted (e.g., 35 mins), and after an hour changes to a time (e.g., 12:20). The 'status' column gives simple information about the users' (apparent) circumstances. The badge information is made available to all computer screens in the organization.

By pressing the badge button, a user can explicitly indicate that he does not wish to be disturbed (busy). This is cancelled automatically when the user moves away from the current cell. It is also possible for the user to specify a phrase to be displayed in the status column (e.g., Away in Italy). Thus when people go away they can indicate this to others. The message is cancelled automatically when they are sighted again. Using automatic rules to cancel features is important because a principle of the design is to require the user to do very little to make it work. To generate location information wearing the badge is all that is necessary. With the 'away' feature an incorrect list of future absences does not matter.

Various versions of the user interface have been tried which include map displays, simple natural text generation systems and others. The Active Badge has also been used in a security role for example to control door locks and to blank a computer screen when the user walks away.

16.3.2 Use of the active badge system

The system has been in use for several years and continues to be very popular. Users interrogating the system have made it one of the biggest consumers of CPU cycles in an organization which uses computers for many applications. Figure 16.4 shows the proportion of time that people in the organization spend in their office and in the building. It can be seen that in this organization people spend a large proportion of their time away from their normal office and one of the reasons why the badges are successful is because less time is spent trying to find others. The most important single piece of information is that somebody is *not* available. A more subtle but equally important observation is that the badges provide a mechanism for people to be less intrusive with respect to others. Our initial thoughts were that the number of phone calls would go up because people could make contact more easily. In fact a self-filtering process

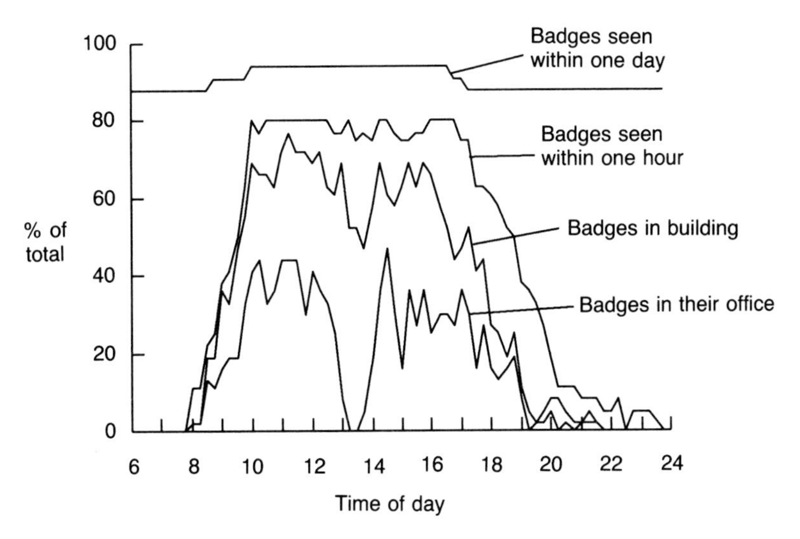

Figure 16.4 Mobility of badges.

takes place where people do not contact others if they see them in circumstances where they may not wish to be disturbed.

Having been first established on a site with 40 wearers (Olivetti Research), the system was extended to a second site with about 80 more wearers (University of Cambridge Computer Laboratory). Because the amount of data generated by movement of people is small, the speed of the communications channel required was not high. Within one organization, the wearing of a badge indicates the user is prepared to have information about their location distributed throughout that organization. Between the two organizations the opposite view is taken and it is necessary for each user to authorize explicitly their sightings being sent elsewhere. The two groups need to communicate frequently and in practice all those working on joint projects (and others) have authorized the information to flow in this way.

Other organizations which work collaboratively are now having the system installed (for example DEC Research). It will be interesting to see if inter-working across continents is also made easier using this approach.

The Active Badge location information has been integrated with the Pandora system. A video-phone can be routed to the Pandora's Box nearest to a user. Who is calling (not where they are) is indicated to the recipient which makes it easier to decide which response to give. It is also possible for both sides of a conversation to know who else is nearby to prevent confidential discussions taking place when other people are in ear shot but out of view. The badge data is also integrated with the video-mail system. Video-mail is most popular in a form where the user does not type anything and there is very little information to help subsequent retrieval. The badge data is used to annotate automatically the video stream. This permits the retrieval of all the video-mail files sent by a particular user, the ones that were recorded with visitors present, or any other category which can be derived from the badge annotations.

16.3.3 Extending desktop information

The cellular infra-red network can be extended to two-way communication which makes it possible for more sophisticated badges to be developed. However, the challenge is to design the simplest systems that provide useful features with only minimal or easy to learn interactions.

Two-way authenticated badges

A new badge has been developed which both receives and transmits. It is implemented using a normal microprocessor and its properties are controlled by software. The two-way badge has a tone generator and two

LEDs making paging applications possible. The paging can be made location dependent so that for example a page as a reminder of a meeting only goes off if the user is not in the meeting room. The Authenticated Badge also has a 'yes' button and a 'no' button which the user can press to generate context dependent commands.

A transmit/receive active badge enables a more secure system to be devised by using a one-way authentication function. The badge identifies itself in the normal way but can also perform authentication by one-way mapping a 48-bit random number to another on demand. For more security it would be possible to incorporate a personal identification number which when revoked disables the badge.

Finally the Authenticated Badge has a simple radio receiver which enables a second type of location system to be implemented. A radio field can be associated with a piece of equipment. When a badge enters this field it is possible to determine within which one (if any) it is. The field can be made small so that when a user approaches a piece of equipment sufficiently close to use, this fact is made known to the system and the appropriate application invoked. The radio field is also used to trigger the badge when it is in low power mode between transmissions.

Badges for equipment

A badge suitable for attaching to equipment is being designed and has similar properties to the transmit-only Active Badge. It uses less power by only transmitting every five minutes and it can be attached in a way which indicates whether the equipment is switched on. By using the network of receivers, some objects can be located at the same level of granularity as personnel.

It is now possible to have an automatic inventory control system which keeps account of what there is and where it is. Tagging in this way will make possible an environment in which the location of all equipment, all I/O devices, all cameras and microphones, is known. With such an active environment it is possible to think of new applications which combine location information of personnel and of equipment.

16.4 THE ACTIVE OFFICE ENVIRONMENT

This paper has described how new communications facilities may be accepted by users. In a discussion entitled 'Technology in the Third Millennium' it is appropriate to speculate about future directions. The 'active office environment' is a framework for development based on experience with the Pandora and Active Badge projects.

16.4.1 Components making up the active office environment

The 'active office environment' will make widespread use of ATM networks. Fibre will be used both for long distance public systems and local private systems, while infra-red and radio will be used for portable applications. Wire drop cables will continue to be available for short distances because they will not have to operate at the same speeds as the underlying networks. Islands of high interconnection will be joined by high capacity links while elsewhere wide area radio and satellite systems will be used to provide some connectivity (I.E.E.E., 1991*b*; 1991*c*). A significant part of the traffic will be bulk image data which can be carried on conventional rather than ATM networks.

The devices using these communications system will be constrained by the facilities they provide and by size and convenience of use. They will all be capable of manipulating audio and video in the same way as any other type of data. A convergence of the way communications are handled within a device and on a network will take place so that there will be similar protocols and little mismatch for use of desktop or remote systems (Hayter and McAuley, 1991).

Automatic location of components will be used to offer users choices that are more comprehensive and easy. The Equipment Badge provides a particular level of granularity for doing this. However, other levels of location granularity may be available as well. Fine grain distribution of location information will need sophisticated algorithms for optimizing traffic flows. Systems which require location information infrequently will make enquiries on a need to know basis. Systems which require location information all the time, or cannot tolerate the delay associated with explicit enquiries, will require such information to be distributed automatically.

16.4.2 Personalization and mobility

A major difficulty when using computer and communications equipment is that user interfaces are not consistent. In the active office environment, the desktop will be made to follow the user and each piece of equipment will be personalized when used. One way this can be achieved is by using a badge. When a user approaches a piece of equipment he is associated with that equipment and the appropriate interface is invoked. To make this facility eventually possible on a world wide basis the Authenticated Badge transmits to the equipment a pointer (similar to an e-mail address) indicating the home base of the badge wearer. If the communications channel is of low bandwidth then the parameters which follow the user would only be of a parametric type. If the communications channel has high bandwidth then all data can follow. An alternative and perhaps complementary way of providing the personalization facility would be to store the parameters describing the users preferences within the badge. An example of personalization in use would make the short codes on every telephone remain the same so that no matter where in the world the call was being made the same destination would be dialled.

Personalization can also be used to control the operation of the communications environment by incorporating the users wishes as specified in his user service profile. This will control distribution of location information, routing and forwarding of calls, and interactions with computer systems. Table 16.4 shows what might be a typical service profile. In this example the user has allowed location and status information to flow to one other organization and also to one individual no matter where that individual is. Call preferences are for the best quality available with the user choosing automatic forwarding and some automatic redirection. Simple interaction with the computer environment is allowed to indicate when video-mail arrives and when the user is in the wrong place with respect to an active calendar. Table 16.4 illustrates a text interface for making these choices as it may be that natural language systems operate well in this restricted domain.

Table 16.4 User service profile

Option	*Choice*
Distribution of location/status information	
Own organization	Yes
Other organizations	Computer Laboratory
Other individuals	No except Fred
Call control	
Video, Audio or Page	Best available
Fixed or portable	Fixed if nearby otherwise portable
Follow-me forwarding	Not if with the boss
	Not if in the library
	Not if in a meeting of 3 or more people
	Not if busy pressed on badge
Redirecting	To secretary in the morning
	Play video message 1 in the afternoon
	Play video message 2 if Fred
Computer Systems Interactions	
Page	Yes if video-mail arrives
	Yes if calendar exception

16.4.3 Communicating in the active office environment

A user will be able to identify the destination as a place, a person, or a facility. He will then choose the appropriate audio, video, or if real-time communication is not required, multi-media e-mail system. To do this, directory services will be provided which will deal with the recipients' mobility and preferences. The distribution of information will be subject to a hierarchy of access restrictions as specified in the user service profile (I.E.E.E., 1992).

There is a tension between the flexibility the receiver may wish to have by receiving all data and the security required by the source. If the caller has access to enough information about the recipient he may know when *not* to make the call and both parties may find this more appealing. When designing such systems it is a challenge to provide security features and methods of use which the users accept, so that information for making such decisions is generally available.

Directory services

Because video, audio and text will be available, one purpose of the directory services will be to provide information about the equipment at the destination and its compatibility with the source. A simple directory service of this type can be provided by relatively infrequently updated tables. A partially dynamic system will require frequent updates. This will enable the potential recipient to indicate status information such as away, portable not in use, or some other choice. A completely dynamic system would present the full picture to the caller including location of the recipient, of equipment nearby, and any other relevant information. This will require the directory to be completely dynamic, that is much of the data would have a short lifetime and have to be updated frequently. The operation of directories and the choices they give callers will be automatic. The implementation of the directory service would take into account the frequency at which entries change and how often they are used.

Call control

When the call is made it may be redirected as specified in the user service profile. If it is routed to the recipient there will be an indication of who is calling. A number of choices for dealing with the call will be available such as accepting the call, giving an immediate response to the caller (e.g., I will call you back), or using the redirection options indicated in the user service profile. If the call proceeds the specified media will be used.

If the recipient is not presented with the call or chooses to reject it an automatic system for redirection will be used. The caller may be redirected in a number of ways or have a text, audio or video indication of what happened. In the example shown in Table 16.4 the recipient has specified that the secretary is to handle rerouted calls in the morning, a video message is to be played in the afternoon, and a special video message is to be played if a particular party calls.

16.4.4 Conclusion

As technologies develop the power of communicating devices will increase. They will also be able to communicate much better. The desktop will no longer be associated with a physical place nor an interface to a single system. It will become virtual and there will be one seamless interface which is consistent across devices and communication systems.

ACKNOWLEDGEMENT

My thanks go to the many colleagues both at Olivetti Research and the University of Cambridge Computer Laboratory whose joint efforts have made much of the work described in this paper possible. I am also grateful to Prof. Maurice Wilkes who made pertinent comments about this manuscript.

REFERENCES

Greaves, D.J., Lioupis, D. and Hopper, A. (1990) The Cambridge backbone ring. *Proc. I.E.E.E. Infocom'90*, San Francisco, June.
Hayter, M. and McAuley, D. (1991) The desk area network. *ACM Oper. Sys. Rev.*, October.
Hopper, A. (1990) Pandora – an experimental system for multi-media applications. *ACM Oper. Sys. Rev.*, April.
Hopper, A. (1991) Design and use of high-speed networks in multi-media applications, in *Proc. 3rd IFIP Conf. on High-Speed Networking*, Elsevier Science Publishers, March.
I.E.E.E. (1990) Multi-media communications, *I.E.E.E. J. Sel. Areas in Comm.*, April.
I.E.E.E. (1991*a*) The promise of the next decade, *Computer*, I.E.E.E. Comp. Soc., September.
I.E.E.E. (1991*b*) Satellite and terrestrial systems and services for travellers, *I.E.E.E. Comms. Mag.*, November.
I.E.E.E. (1991*c*) Architectures and protocols for integrated broadband switching, *I.E.E.E. J. Sel. Areas in Comm.*, December.
I.E.E.E. (1992) Intelligent networks, *I.E.E.E. Comms. Mag.*, February.
Want, R., Hopper, A., Falcao, V. and Gibbons, J. (1992) The Active Badge location system, *ACM Trans. on Inf. Sys.*, January.
Want, R. and Hopper, A. (1992) Active Badges and personal interactive computing objects, *I.E.E.E. Trans. on Consumer Elect.*, February.

Index

Page numbers for figures are shown in bold; page numbers for tables are shown in italics.